Juergen Alt

The Small Book about Design-for-Test

THE SMALL BOOK ABOUT DESIGN-FOR-TEST

A comprehensive guide to Design-for-Test methods in the semiconductor industry by Juergen Alt

Bibliografische Information der Deutschen Nationalbibliothek: Die Deutsche Nationalbibliothek verzeichnet diese Publikation in der Deutschen Nationalbibliografie; detaillierte bibliografische Daten sind im Internet über dnb.dnb.de abrufbar.

Verlag: BoD · Books on Demand GmbH, Überseering 33, 22297 Hamburg, bod@bod.de

Druck: Libri Plureos GmbH, Friedensallee 273, 22763 Hamburg

ISBN: 978-3-8192-7665-1

Foreword

This book is designed for individuals working in the semiconductor industry who need a fundamental understanding of Design-for-Test (DfT) methods. It caters to test and product engineers who typically utilize these methods, as well as chip designers, project managers, and business owners who seek to comprehend the value of implemented DfT methods. Additionally, it serves as a starting point for beginners in DfT before delving deeper into implementation tools and test pattern generation tasks.

Originating from a university lecture, the core content of this book has expanded over the years. It bridges the gap between theoretical textbook descriptions of DfT and its practical application in the industry. The goal is to present DfT as a collection of methods that maintain manufacturing test costs at a level that allows for reasonable selling prices, even for complex products utilizing the latest silicon technologies.

The book is structured into three parts:

- The first part provides a summary of DfT techniques and an introduction to pattern generation techniques.

- The second part delves into established DfT techniques for special analog-mixed signal circuits and memories within the industry.

- The third part proposes a systematic approach to test concept engineering, maximizing the benefits of DfT methodologies.

The author, with over 30 years of experience in the semiconductor industry, began his career as a DfT expert following his PhD. Throughout his professional journey, he has collaborated closely with design, test, and product engineering teams in various roles. For the past decade, the author has been a lecturer at FAU Erlangen-Nuremberg, Germany, empowering the next generation to continue innovating in test and DfT. One of the key insights from the author's career is the continuous need for innovation across all fields of semiconductor development. This book provides a snapshot of DfT as of 2025, with the expectation that future chapters will be written by others.

By exploring the fundamentals and advanced concepts of DfT, this book aims to equip readers with the knowledge and tools necessary to contribute to the ongoing innovation in semiconductor testing and development.

Contents

DfT to optimize manufacturing test

In semiconductor development, two critical disciplines harmonize to ensure reliable products: Design and Manufacturing Test. The Design team's primary responsibility is to create product functionality based on a specified set of requirements. In contrast, the Manufacturing Test team's role is to verify that each manufactured device conforms to the design's expectations.

It is essential to note that the purpose of manufacturing tests is not to uncover any functional design flaws. Instead, these tests focus on verifying the quality and correctness of the physical manufacturing process. Given the inherent nature of wafer manufacturing and die assembly, there exists a probability of defects in the final devices.

Ideally, the most effective way to validate a manufactured device would be to run it through the intended customer applications and all relevant use cases. However, this approach is typically impractical for almost all semiconductor products. The reasons for this include:

- Not all use cases are known beforehand.

- Products operate across a wide range of voltages and temperatures.

- There is a high degree of programmability involved.

- Exhaustively testing even small digital blocks would require an immense amount of time due to the vast number of possible input combinations and outputs that need validation.

Despite these challenges, an exhaustive test approach, where feasible, remains the gold standard for ensuring product quality and reliability.

Test equipment interaction with device-under-test

The ATE and DUT interaction model shows the main components of Automated Test Equipment (ATE). These are:

- Data Source
- Evaluate Responses
- Clock Generation
- Power Supply
- Analog and RF Measurements

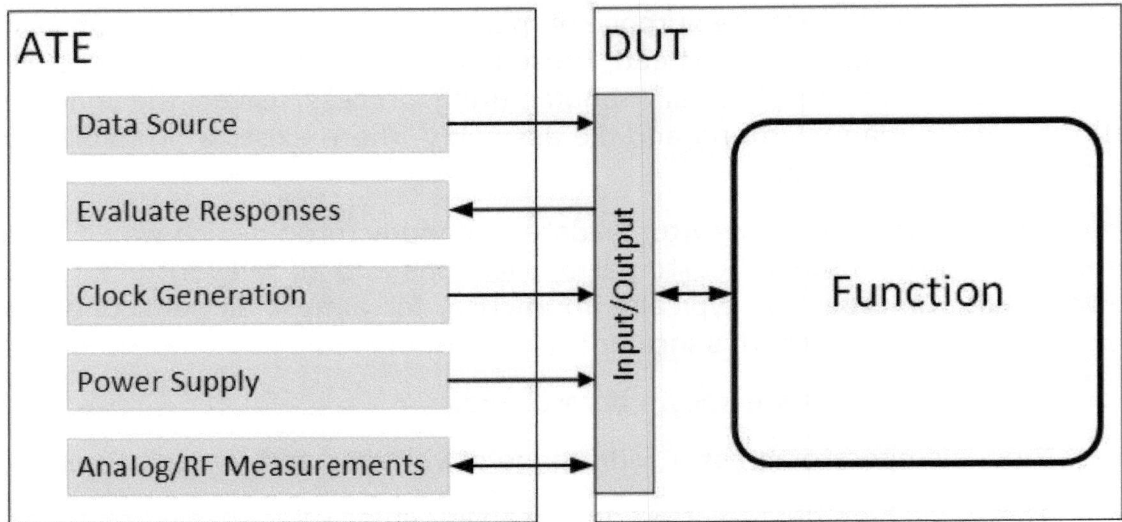

Figure 1: ATE_DUT interaction model

Analog and RF measurement capabilities are optional. Depending on the product to be tested different features are needed. The required measurement capabilities are defining the costs for ATE. Other significant cost drivers are possible test frequencies, memory depths for data sources and expected responses. The number of channels available also influences the ATE costs.

These channels are the communication interface to the device-under-test (DUT) or circuit-under-test (CUT) shown on the right side of the graphic. The DUT is only accessible via the input and output structures of the device. These structures are pad patches for dies on wafer. Depending on package type, pins or balls are the points of access for a packaged device.

ATE are general purpose machines, i.e., they can be used for different kind of products. The adaptation between ATE and product is established using:

- Probe cards for testing dies on wafers
- Load boards for testing packaged devices

Manufacturing test at ATE is controlled via a test program. A test program applies a set of single tests. Generally the tests are categorized into three groups:

- Parametric
- Functional
- Structural

A single test is called a pattern. With every pattern a timing for the tester channels and power settings are associated.

Parametric tests confirm that data sheet parameters of the device are within specified limits. Examples are current consumption or parameters for analog interfaces like Analog-to-Digital Converters (ADC).

Functional patterns assume a black-box model for the test. The behavior of the tested function is checked by stimulating the inputs and checking the expected behavior at the outputs. The complexity for a complete functional test increases significantly with the number of function's inputs and internal states. Assume $n=50$ internal states and $k=25$ inputs for a sequential circuit according to Figure 2. The exhaustive test set for this circuit has a size of 2^{25+50}. If the test is executed with a frequency of 100 MHz the test would take 12 million years.

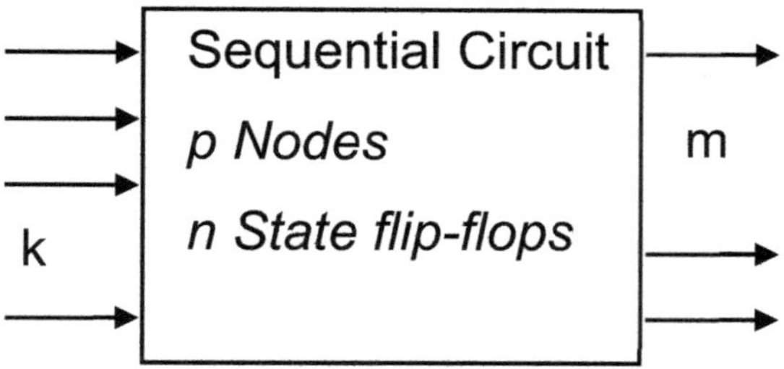

Figure 2: Sequential circuit

By applying functional tests no information is available about the root cause of a defect. There is also a possibility that a fault effect is blocked internally and not visible at the outputs.

Structural patterns are constructed using the internal realization ('structure') of a circuit. The function is out of scope for structural test. Structural tests are developed using the structural realization, as for example a gate level netlist. Defects or faults are inserted into this netlist which should be detected by the applied test sequences. Structural pattern development is closely linked to DfT measures implemented into the design.

Test time, yield, quality and reliability

Test and DfT contribute to achieving a product's business targets. Targets for test time, yield and area for DfT are directly linked to manufacturing costs or 'Cost of Sales' (CoS). Development efforts for test program and DfT implementation are part of research and development (R&D) cost. A typical profit and less statement is shown in Figure 3.

Revenue is what you get on the market for your product. The average selling price (ASP) is multiplied with the number of products sold. This revenue is required to cover certain types of costs:

- G&A: These are general and administration costs to finance central functions of a company as human resources or legal departments.
- Sales: These costs cover expenses for sales activities.
- R&D: Research and development activities focus on developing new products. Most of these costs are related to the engineering work force. R&D costs have to be spent independently of the manufactured volume of a product.
- CoS: Cost of Sales cover all spending to produce goods sold. Compared to R&D cost this cost type depends on manufactured volume. Basically, there is zero CoS if the product is not manufactured.

For a healthy business case there should be a (positive) profit. Subtracting all 4 cost types listed above from revenue gives the achieved profit.

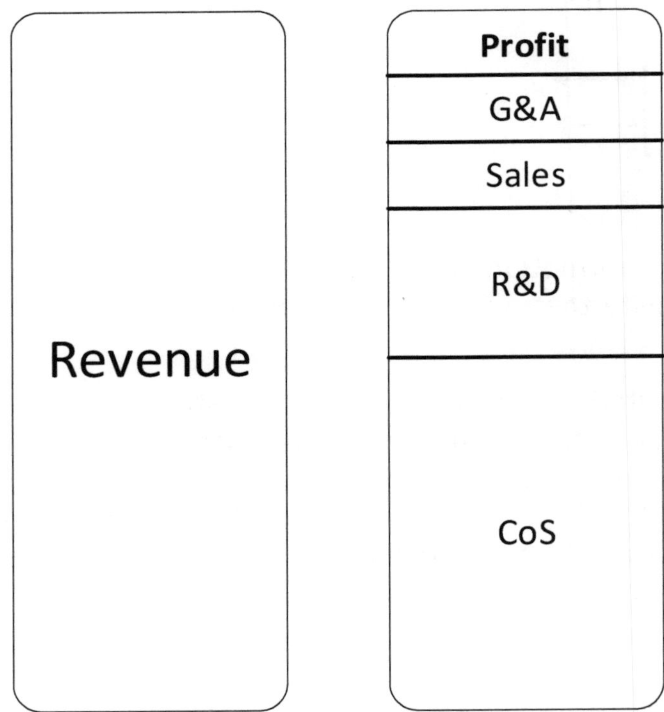

Figure 3: Profit and loss statement

A closer look into CoS gives the following elements:

- Cost for wafer manufacturing (FE cost): Wafer cost charged by foundry or internal wafer fab depends on technology node and process complexity. Costs are higher for more advanced technology nodes, i.e., smaller geometries. A higher number of metal layers increases the process complexity resulting in higher costs, too. Breaking down cost to die level requires consideration of the achieved Yield.
- Cost for package and assembly (BE cost): All cost types for backend manufacturing are covered here. The main contribution is the type of package used for the product.
- Cost of test: Typically, there are different test insertions associated with wafer manufacturing as well as with package and assembly. For the following we summarize cost for frontend test insertions (FE test) and for backend insertions (BE test).

Test costs consider the investment of the required ATE, their operational expenses and the time a product uses the test equipment. Investment and operational expenses are typically covered by the ATE Cost Rate[1]. To get test costs for a test insertion this rate is multiplied by the time needed to run all patterns for this insertion. Below, test time is replaced by its reciprocal 'throughput'.

- Test cost = ATE Cost Rate * 1/Throughput

- Throughput: Number of parts tested each time unit (1/Test Time)

- Cost Rate: ATE cost ($) for each time unit

Yield

For considering yield at wafer manufacturing, we assume randomly distributed defects all over the wafer. If 1000 identical dies are manufactured on a single wafer, a perfect manufacturing test would identify all dies as failing ones which cover one or more defects. If 50 dies are failing the resulting yield would be 95%.

- Yield = Pass parts / All parts

Now we assume the identical distribution of defects on the wafer, and we manufacture bigger dies resulting in 500 identical pieces for a single wafer. The probability that a die covers one or more defects is higher now and the resulting yield is lower. If otherwise smaller dies are manufactured assuming the same defect distribution yield will increase. To conclude these thoughts, yield

---

[1] Usually given as $/h or €/h for a complete test cell including associated equipment like handler or temperature control unit.

increases with decreasing die area. Additional to area the Price[2]-Formula [1] gives other basic parameters influencing yield:

$$Y = \frac{1}{(1 + A \cdot D_0)^n}$$

In this formula A is the area of a die, D_0 is the defect density per manufacturing layer and n is the process complexity factor. Figure 4 shows examples for different die sizes using the same defect distribution on wafer indicated by red dots.

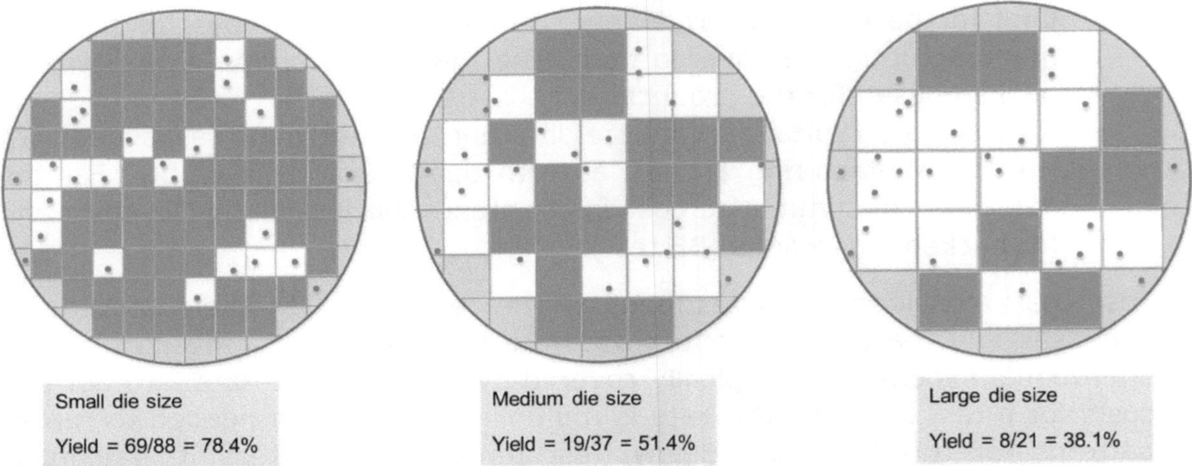

Figure 4: Yield for different die sizes

With every processing step there is a yield loss as shown in Figure 5. Yield is defined generally as 'good pieces out' divided by 'pieces in'.

---

[2] Named after John Price who proposed this formula.

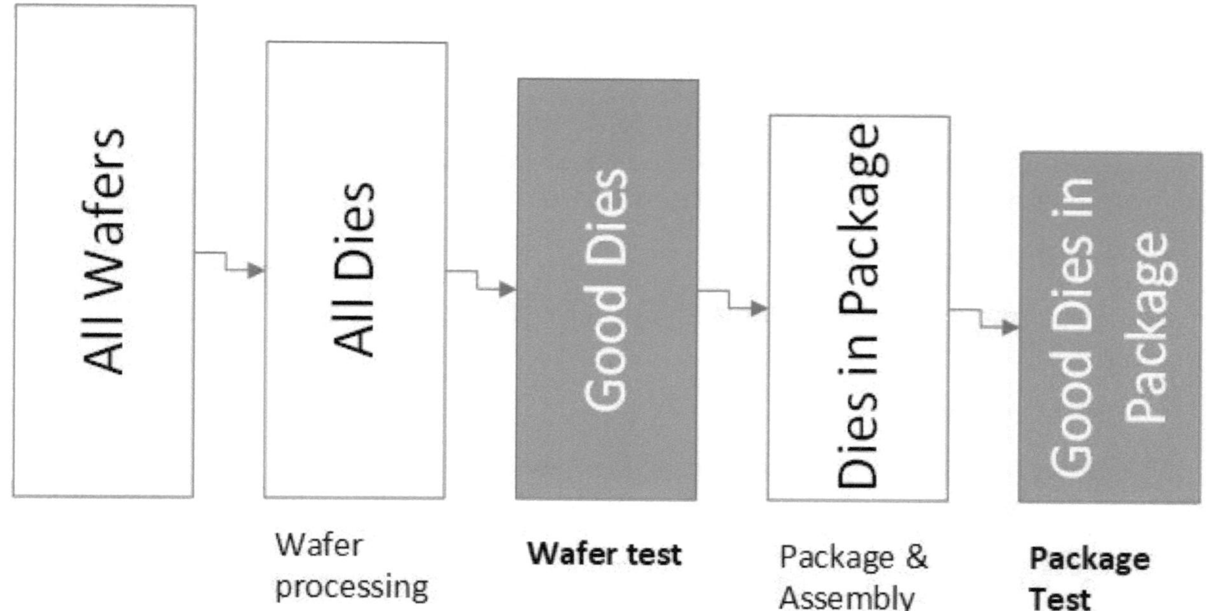

Figure 5: Yield loss identified by wafer and package test

Unwanted yield loss during test insertions is related to 'over-testing' and to instable test patterns. Both originate from operating points for patterns like voltage, timing and power which are at the edge of the device's specification.

Quality and Reliability

Not directly visible within profit and loss statement are business objectives like quality and reliability. Besides cost targets every product has to fulfill a set of requirements [2]. For a semiconductor product examples for requirements are function, performance of processing units, specific interface parameters but also general parameters like operating temperature range, operational voltages and power consumption. A product's specification documents all intended functions or features for customers. Semiconductor companies have to guarantee that all manufactured products behave according to their specification. The main purpose of manufacturing test is to guarantee the so called 'shipped-quality' towards the customers. A measurable for the shipped-quality is the defect level or fail rate of products after delivery. This fail rate is measured in defective-parts-per-million (dppm). The expectations on dppm rate depend on application and market segment. The quality requirement (i.e., low dppm rate) is highest for safety, security and high availability applications as dominant in automotive, aerospace and military markets.

The first approach to model the fail rate of semiconductor devices dates back into 1981 [3]. The modeling is based on probabilities:

- Probability for a fault is p_f, assuming that all faults are equally likely

- Probability of a chip being good is $(1-p_f)^f$, where f is the number of all possible faults

The last term is the representation for yield. The probability that a fault is not detected by the test is assumed as escape probability p_e. Using the terms and probabilities introduced a fail rate or defect level can be defined as:

$$DL = p_e / [(1-p_f)^f + p_e]$$

If we assume that a test is able to detect m faults with $m \leq f$ the following can be formulated:

$$(1-p_f)^m = (1-p_f)^f + p_e$$

The term for the defect level can now be written as:

$$DL = 1 - (1-p_f)^{f-m} = 1 - Y^{(1-m/f)}$$

Here $Y = (1-p_f)^f$ is the yield and m/f is the coverage of the applied test.

The model includes no differentiation for structures like embedded memories, analog, logic or others. Also, a single coverage is used here without considering different fault models. This model is able to show general relations between yield of the manufacturing process, coverage of the applied tests and the resulting shipped-quality.

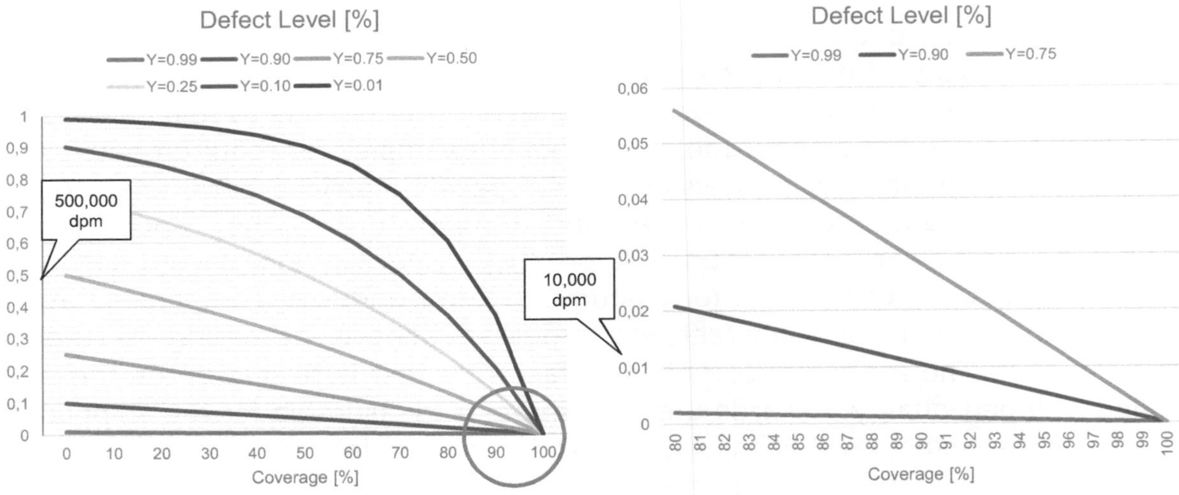

Figure 6: Fail rate or defect level as function of coverage

Figure 6 plots the formula for defect level and coverage using different yields as parameter. The coverage range from 80% to 100% is shown enlarged on the right side. The effect of manufacturing yield on failure rate can be seen clearly. High coverages with more than 95% are required for yield of 90% to achieve failure rates below 500 dppm. For automotive quality ranges below 1 dppm coverages at >99% level are required.

Sometimes quality is mentioned as '0h-quality' to differentiate clearly from reliability. Reliability is quality over lifetime and shown as a 'bathtub-curve' in Figure 7.

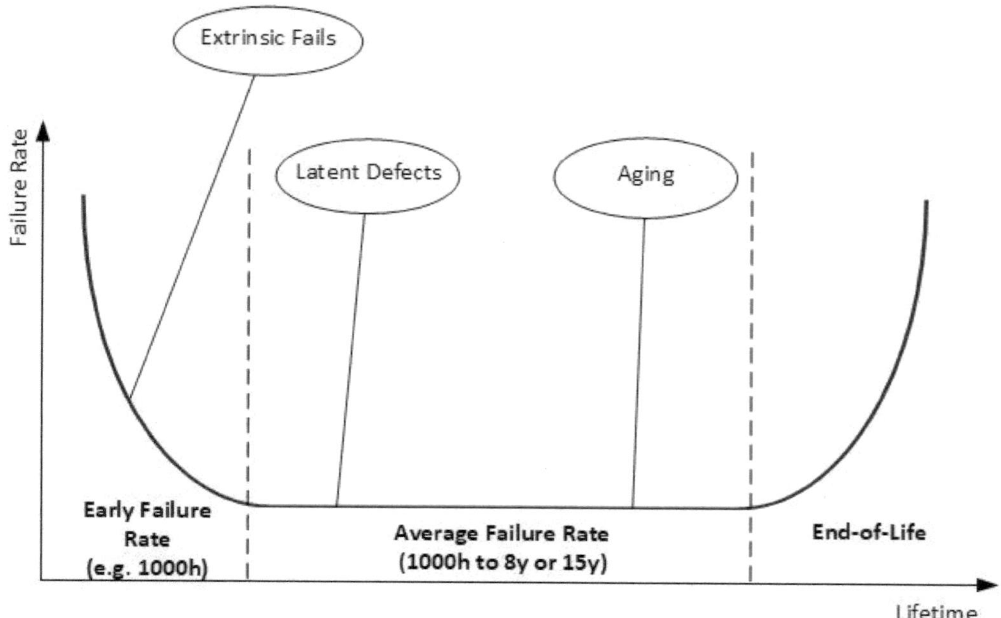

Figure 7: Failure rate over lifetime ("bathtub curve")

Typically, a product's failure rate is high during first 1000h of operation. An average, lower failure rate is achieved during normal lifetime period. This period depends on application and market segment. Again, consumer products including mobile devices have lower lifetime requirements than automotive products. At the end of lifetime the failure rate is increasing again. Lifetime is influenced by use cases for the application. Hours of operation, stand-by and switch-off periods combine into a use case and determine possible lifetime. Reliability requirements are formulated as FIT rates (failures-in-time)[3]. The FIT rate measures the number of failing devices per 10^9 operating hours[4]. For this an assumed use case is required. E.g., operation time for a vehicle's engine depends if a truck is on the road for above 12h per day or a privately owned car which is used for 1-2h per day only.

For high reliability products stress tests are part of the manufacturing test flow. Objective of stress tests is to activate extrinsic faults before the product is shipped to the customer and used in targeted application. The product's lifetime starts with the average failure rate according to Figure 7.

---

[3] German: Ausfallrate

[4] Sometimes other number for time interval used, should be given to avoid confusion

Fault models

All fault models of practical use are described on gate level. This abstraction level represents a realization using Boolean gates taken from a standard cell library and their connections to implement the circuit's specified function. Typically, the gate level netlist is generated by the synthesis process. As shown in Figure 8 the synthesis software uses an RTL description as input and map it to the targeted manufacturing technology using a standard cell library ('Design Library'). Such a design library contains cell descriptions for simple Boolean gates realizing AND, NAND, OR and other basic functions for a specific manufacturing technology [35].

```
1 library ieee;
2 use ieee.std_logic_1164.all;
3 use ieee.numeric_std.all;
4
5 entity signed_adder is
6   port
7   (
8     aclr : in    std_logic;
9     clk  : in    std_logic;
10    a    : in    std_logic_vector;
11    b    : in    std_logic_vector;
12    q    : out   std_logic_vector
13  );
14 end signed_adder;
15
16 architecture signed_adder_arch of signed_adder is
17   signal q_s : signed(a'high+1 downto 0); -- extra bit wide
18
19 begin -- architecture
20   assert(a'length >= b'length)
21     report "Port A must be the longer vector if different sizes!"
22     severity FAILURE;
23   q <= std_logic_vector(q_s);
24
25   adding_proc:
26   process (aclr, clk)
27     begin
28       if (aclr = '1') then
29         q_s <= (others => '0');
30       elsif rising_edge(clk) then
31         q_s <= ('0'&signed(a)) + ('0'&signed(b));
32       end if; -- clk'd
33     end process;
34
35 end signed_adder_arch;
```

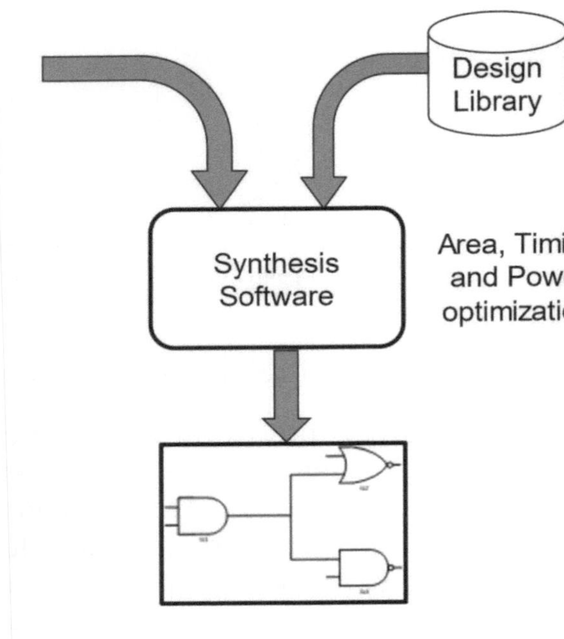

Figure 8: Synthesis process from RTL to gate level

Also, more complex gates are included, e.g. multiplexer or full-adder. For a gate function multiple variants are available. Such variants cover small area cells with low driving capability and large area cells with higher driving strength. Those variants allow the synthesis software to optimize area, timing and power as specified by the user for the circuit.

Stuck-at faults

The stuck-at fault model uses internal nodes between gates as fault site or fault location, see Figure 9. Two fault types are assumed for every fault site:

- Stuck-at ‚0' fault: internal node is at logical level ‚0' for complete test period (connected to GND)

- Stuck-at ‚1' fault: internal node is at logical level ‚1' for complete test period (connected to VDD)

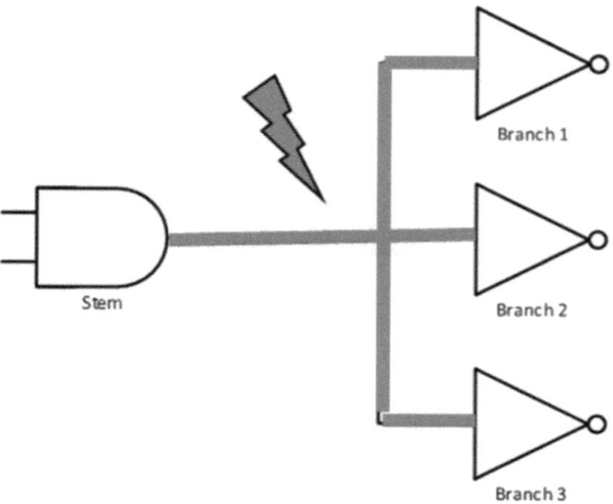

Figure 9: Fault sites (locations) for stuck-at faults

The stuck-at model is a static fault model since it is based on the assumption that a fault is present for the complete test period. The total number of faults using this model is approximately twice the number of internal nodes.

Delay faults

Dynamic fault models are called delay faults. These models assume that a nominal delay for a signal transition at the fault site is delayed, see Figure 10 (right part). Two fault sites are used for delay faults as shown on the left side of Figure 10:

- Fault site is an internal gate: Transition (or gate) delay fault model
- Fault site is an internal path: Path delay fault model

For both fault models two fault types are used:

- Slow-to-rise: 0->1 transition is delayed
- Slow-to-fail: 1->0 transition is delayed

For transition delay fault model, the total number of faults is twice the number of internal nodes, the same as for the stuck-at fault model. For the path delay fault model, the number of internal paths is dependent on the circuit's functionality and can be significantly higher than twice the number of internal nodes. For practical reasons if path delay faults are used only a subset of faults is selected, typically the longest ones and by this most timing-critical paths are selected.

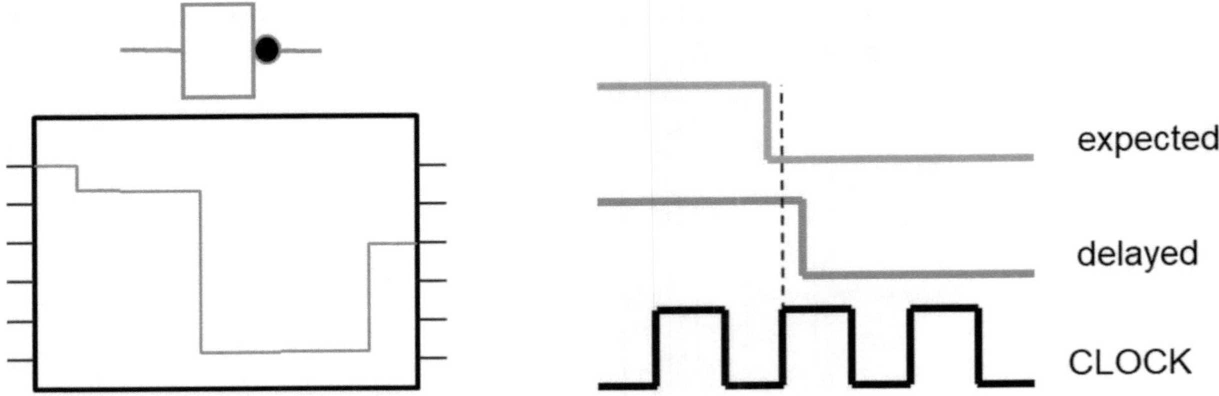

Figure 10: Fault sites and timing for delay faults

Cell-aware faults

The idea of cell-aware faults is to keep the efficiency of gate level circuit for ATPG and fault simulation while accuracy of fault modelling is increased. A first proposal to add a fault description to a standard cell library was presented in [36]. A commercialization of this approach and a close link to an industry standard ATPG tool was introduced by [37]. The following three figures are taken from the last-mentioned publication. To understand the approach a basic understanding of the D-algorithm is required, see Chapter Basic ATPG algorithm and D-calculus.

For all library cells defects inside a concrete layout realization are assumed. Figure 11 gives the layout of a multiplexer cell. The cell output Z as well as inputs D0, D1, D2, S0, S1 are shown.

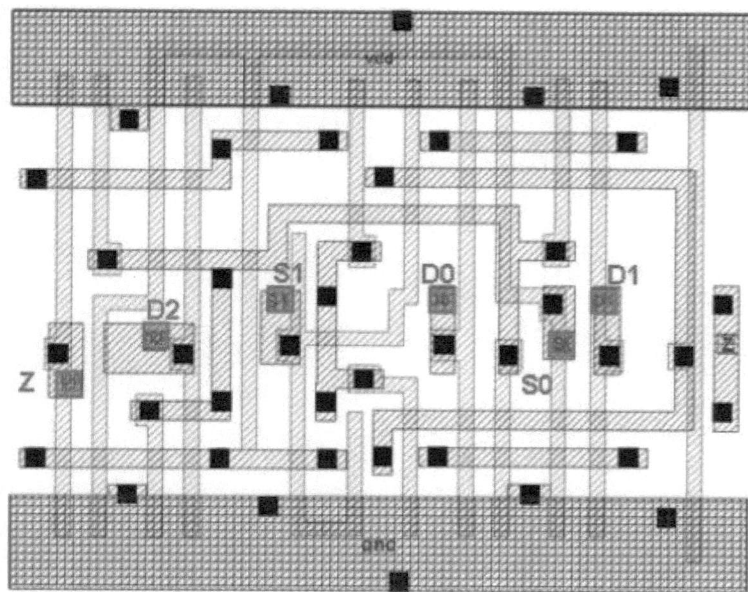

Figure 11: Layout of a multiplexer cell with assumed defects

Transistor level simulations are used to get input vectors for the cell which would detect the internally assumed defects. These detection conditions are stored in a detection matrix for all defects. The detection matrix forms an additional view for every cell. For the multiplexer example the resulting detection matrix for 48 defects is shown in Figure 12. This mapping of cell-internal defects to detection conditions at the cell boundary is a one-time effort for every cell library.

stimuli	d1	d2	d3	d4	...	d41	d42	d43	d44	d45	d46	d47	d48
00000	-	-	-	-		-	-	-	D	-	-	-	-
00001	D	D	-	-		D	D	D	-	D	D	-	-
00010	-	-	-	-		-	-	-	D	-	D	-	-
00011	-	D	-	-		D	-	D	-	D	-	-	-
00100	-	-	-	-		-	-	-	D	D	-	-	-
00101	D	-	-	-		D	D	D	-	-	D	-	-
00110	-	-	-	-		-	-	-	D	D	D	-	-
00111	-	-	-	-		D	-	D	-	-	-	-	-
01000	-	-	-	-		-	-	-	D	-	-	-	D
...													
11111	-	-	-	-		D	-	-	-	-	-	-	-

Figure 12: Detection matrix

An enhanced ATPG tool is able to handle the detection conditions. As shown in Figure 13 the D-frontier starts at output Z of the cell. For a certain defect the stimuli is taken from the detection matrix and gives a condition to be satisfied by the ATPG tool.

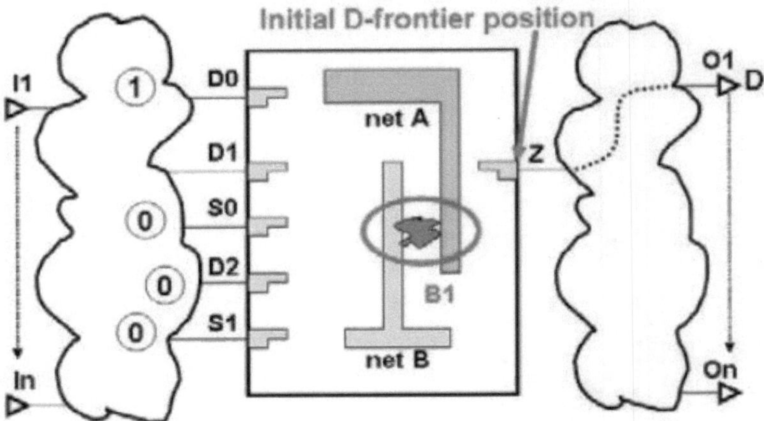

Figure 13: D-frontier starts at cell's output

With this approach the accuracy of fault modelling is increased by a pre-processing step and the efficiency of gate level description can be retained for the pattern generation process. Cell-aware faults can be of stuck-at or transition delay type. There is no additional value for two-input gates since all input combinations are addressed by input and output faults already. The resulting test sets for cell-aware and stuck-at faults would be identical. For cells with more than two inputs additional vectors are added.

Fault coverage definition

For a given fault model and a given test set a fault coverage can be defined. This is:

FC = (Number of detected faults)/(Number of modeled faults) * 100%

A fault coverage is connected to all the following:

- The circuit-under-test
- The used test set
- The underlying fault model

The fault coverage gives an indication about the testability of a circuit and the quality of the test set. It gives no indication about the quality of the fault model itself. There is also no compound fault coverage combining various fault models.

For practical use there is also the term test coverage used. Here all detectable faults are used replacing all modeled faults in the fault coverage definition above:

TC = (Number of detected faults)/(Number of detectable faults) * 100%

Thoughts about fault models and test methods

Generally, models always give a limited view on reality. A good model has just enough information for the task it is needed for. For describing the flight of a ball across a tennis field the model for the ball can be simple by just using the mass concentrated in the ball's center of gravity. To describe how the ball behaves if it touches ground requires more information on ball's shape and material.

The introduced fault models increase in complexity from stuck-at to delay to cell-aware faults. The effort to evaluate the coverages based on the underlying fault models increases with the model's complexity. Nevertheless, tests with 100% coverage for all mentioned fault models do not guarantee that all possible defects are detected. Since a model is always an abstraction of reality there is no 'perfect model' available.

Abstraction levels used in microelectronic design are also used for fault modelling. Fault models defined on the gate level have proven to be effective (i.e., have sufficient accuracy) and efficient, in the sense the EDA tools available for test development have a sufficient performance.

The role of different test methods and fault models is to cover more defects which occur during manufacturing. In Figure 15 the large rectangle represents all possible defects. Kind and number of defects are never known completely. Bubbles in Figure 15 indicate different fault models or test methods. By introducing another test method or applying patterns based on an additional fault model hope is, to cover more of the defect universe than before the introduction.

By increasing the fault coverage of a certain fault model, the size of the bubble in Figure 15 increases. Experiments based on a statistically significant volume of product samples can show the cut set of overlapping fault models and test methods. The uncovered grey area represents the quality risk of a product.

Figure 14 shows the detection of non-target faults or unmodelled defects. It can be expected that a test pattern with a known coverage for stuck-at faults also detects a certain number of other defects which may or may not be detected by other tests.

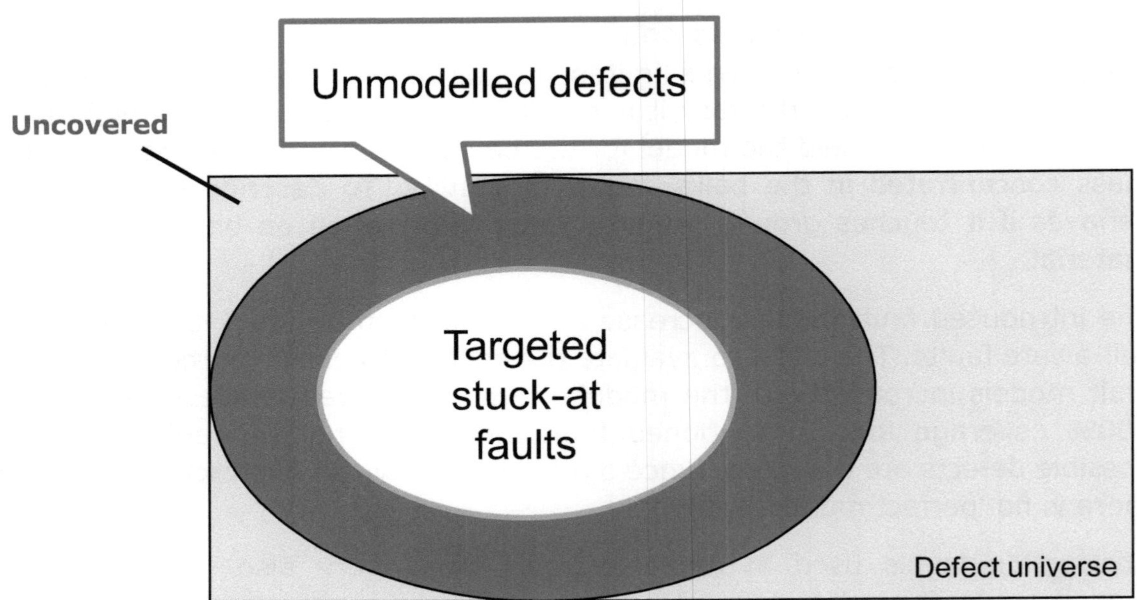

Figure 14: Detection of unmodelled defects

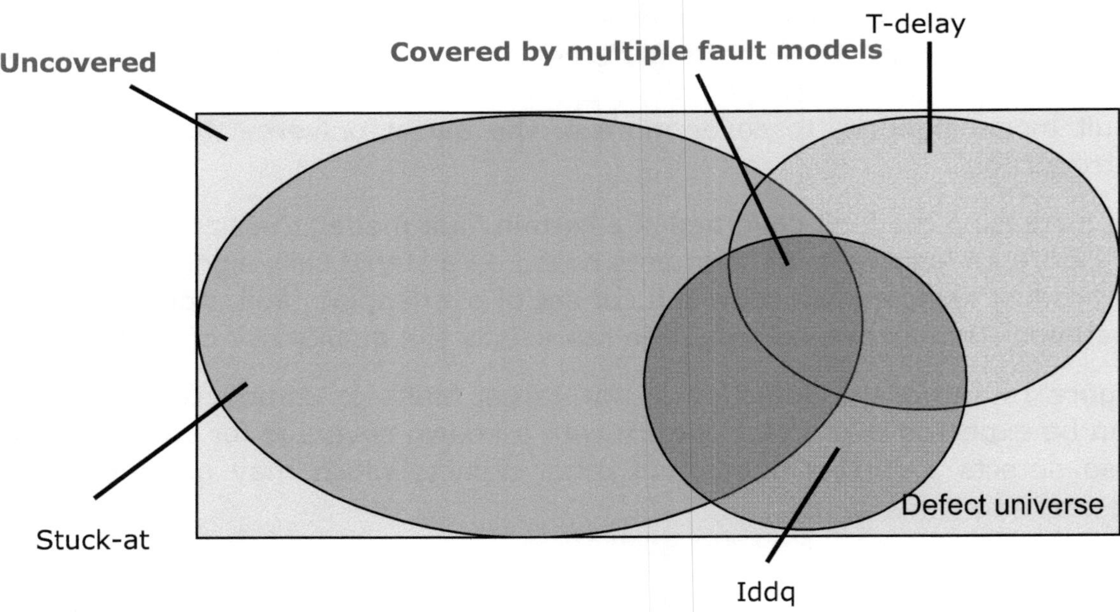

Figure 15: Covering the defect universe

Fault simulation

The principle of fault simulation is very close to the general problem of testing digital circuits as formulated by Edward Moore in 1956 [40]. In this publication Moore defined a machine identification experiment. A machine in this context is a digital automaton containing a finite set of states and combinational circuit to calculate the next state and the machine's output out of the current state and the current input to the circuit. The machine identification experiment should decide if two given machines are identical by input/output experiments only.

The problem statement of fault simulation is to determine the fault coverage for

- Given netlist (circuit)
- Given fault model
- Given test set

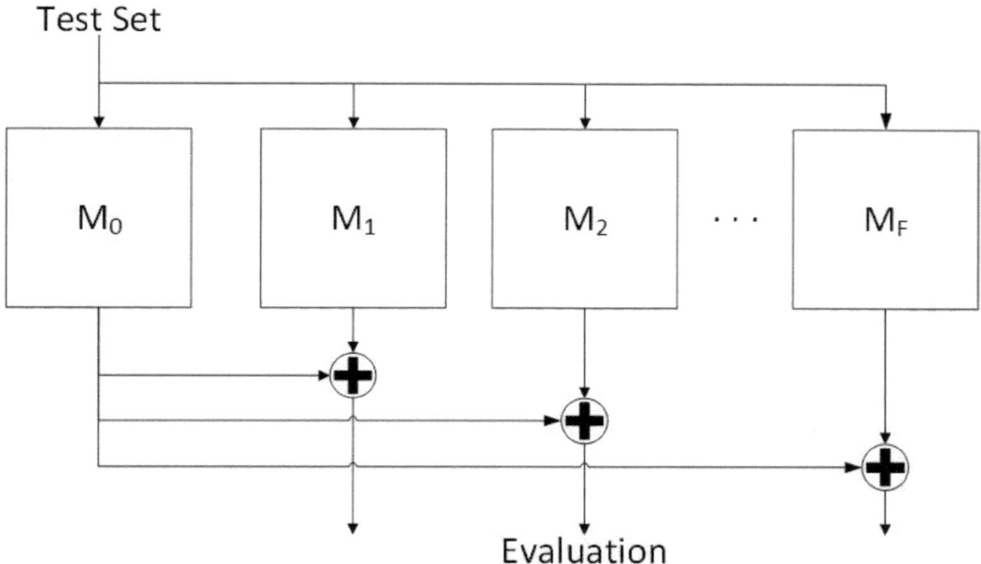

Figure 16: Principle of fault simulation

Mapping this problem statement to the machine identification experiment is shown in Figure 16. There is a known-good machine M_0 which is used as a reference. For each fault of the fault model a faulty machine is created. In total F faulty machines M_1, M_2, ... M_F are available where F is the number of all faults for the selected fault model. The outputs of all faulty machines are compared to the outputs of the fault-free machine M_0. This comparison is indicated by the EXOR gates. The test set is the pattern, i.e., a sequence of test vectors, applied to the inputs of all machines.

Single stuck-at fault example

Taking the single stuck-at fault model as an example the machines are represented by netlists of the circuit-under-test. M_0 netlist is unmodified and fault-free. Netlists for faulty machines are modified at the fault site with the respective value for stuck-at 1 or stuck-at 0 fault. Since two faults are assumed for each internal node (i.e., the fault site) a total of F = (2 * #internal_nodes) faulty machines are created.

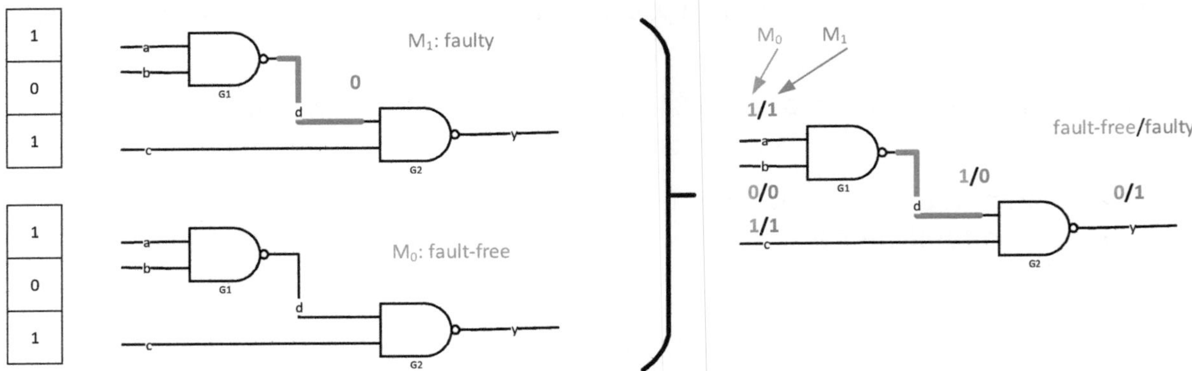

Figure 17: Fault simulation example (1)

In Figure 17 the basic fault simulation approach is shown using a 2-NAND gate circuit with inputs a, b, c and an output y. It is investigated if the input vector a=1, b=0, c=1 detects the stuck-at 0 fault at node d. As a first step the fault site is fixed to d=0. This creates the faulty machine for this specific fault. The fault-free machine remains unmodified, and the test vector is applied to both machines. The combined model on the right shows the logical values for every signal for both the fault-free machine and the faulty machine. For the selected example the fault is detected by the given test vector since a difference between fault-free and faulty machine is observed.

Taking the same test vector and investigating the stuck-at 1 fault at node d is shown in Figure 18. For this fault the test vector is unable to detect the fault since there is no difference at output y for the fault-free and the faulty machine.

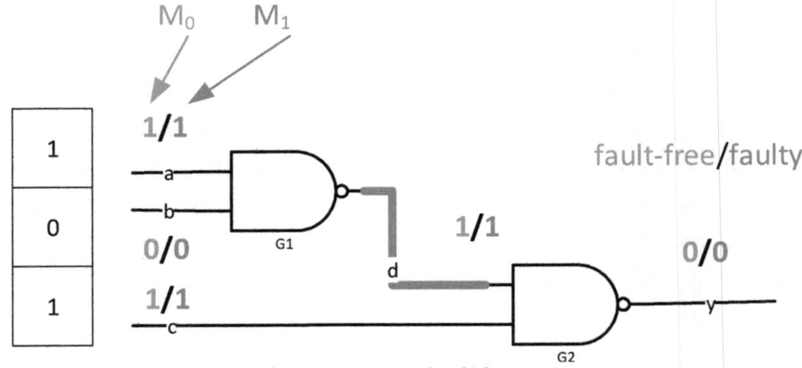

Figure 18: Fault simulation example (2)

Complexity

Time and space complexity are important considerations how scalable a methodology is. Complexity is measured against the size of a problem. Time complexity examines the run time of a process. Space complexity views at the memory (or storage) requirements of a methodology. For fault simulation the abstraction level defines the type of elements to be simulated. Transistors, gates or more abstract blocks like adders have a different accuracy of description and require different compute effort and memory space to store the description. The number of elements to be simulated is defined by the size of the circuit. The selected fault model defines the number of faulty machines to be considered.

To assess complexity single stuck-at faults at gate level are used as a concrete example:

- Circuit size, number of gates: N_{gates}
- Fault model size, number of faults: $N_{faults} \approx 2* N_{gates}$

The effort to simulate a single machine is $N_{gates}* t$, with t as required time to simulate one gate:

- Time for simulating a single vector: $T_{fsim} \approx N_{gates}* t + N_{faults}* N_{gates} * t = t * (N_{gates} + 2N_{gates}^2) => O(N^2)$
- Memory for simulating a single vector: $M_{fsim} = N_{gates} * (1+N_{faults}) = N_{gates} * (1+2*N_{gates}) = N_{gates} + 2N_{gates}^2 => O(N^2)$

For both, time and space a quadratic complexity is inherent to the fault simulation problem. This results in 4x time and memory increase if the circuit size doubles. The complexity of logic simulation ($T_{lsim} = N_{gates} * t$) is just linear.

Effort reduction measures

To reduce the efforts for run time and storing of fault-free and faulty machines some measures are typically used for fault simulation.

The first measure uses the identification of fault equivalence classes. To identify such classes all possible faults for a gate are listed. For a 2-input NAND the possible faults are shown in Figure 19. In total six faults are possible for a 2-input Boolean gate. Therefore, six faulty machines have to be simulated for such a gate.

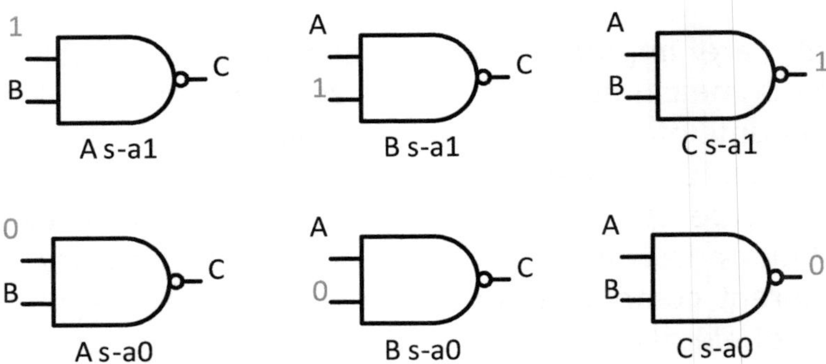

Figure 19: Single stuck-at faults for a NAND-gate

As a theoretical experiment a fault simulation for this gate is executed using all possible input combinations. The result of this fault simulation is listed for the 2-input NAND circuit with following table:

A	B	C	A s-a1	A s-a0	B s-a1	B s-a0	C s-a1	C s-a0
0	0	1	1	1	1	1	1	*0*
0	1	1	*0*	1	1	1	1	*0*
1	0	1	1	1	*0*	1	1	*0*
1	1	0	0	*1*	0	*1*	*1*	0

A specific fault is detected if output C has different value than for the fault-free case. Italic-marked entries show when a vector detects a certain fault. The last vector detects three faults: A s-a0, B s-a0 and C s-a1. These faults have an equivalent behavior since also no other test is able to detect a fault out of this set. Additionally, there is no differentiation possible for equivalent faults. To reduce effort for fault simulation only one of the three faulty machines has to be simulated. This faulty machine represents all three faults.

The second method to reduce effort is called 'fault dropping'. As soon as a fault is detected the faulty machine will be taken out of the simulation process.

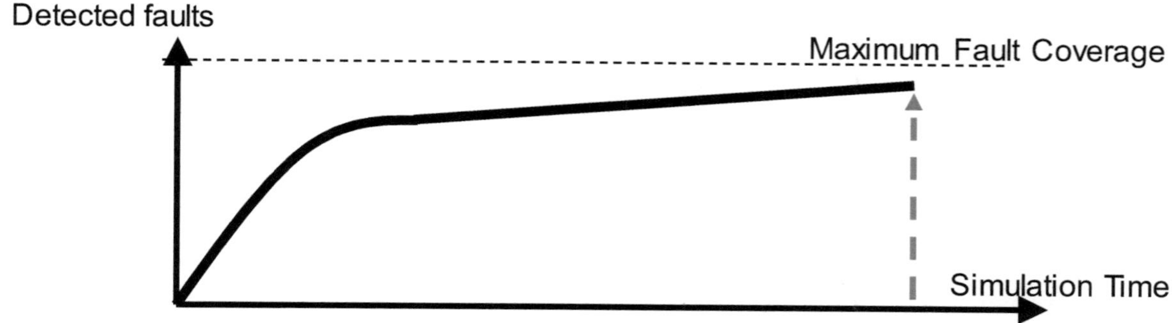

Figure 20: Fault dropping after detection

The basic task of fault simulation is to determine the fault coverage for an applied test set. A typical function for fault coverage over applied vectors (i.e. simulation time) is shown in Figure 20. If fault dropping is applied the faulty machine is simulated until the first vector detects the fault. For the rest of the test sequence the fault is dropped, i.e. the faulty machine is no longer simulated. The area between the maximum fault coverage line and the graph in Figure 20 represents the number of faulty machines to be simulated.

An established method for various simulation tasks is the use of event-driven simulation. Here only the signals are simulated whose values have changed compared to previous state. In Figure 21 the regions with changed values are white-colored and named 'activity cones'. Only those areas are simulated. The other blue areas have unchanged values and the effort to simulate these nodes can be skipped.

Figure 21: Event-driven simulation using activity cones

The algorithm for concurrent fault simulation supports the effort reduction measures presented above, and its idea is close to the machine identification experiment. To explain the algorithm Figure 22 shows three gates as part of a bigger circuit. A logic simulation (machine M_0) initializes all gates within the circuit. In Figure 22 these are gates A, B and C.

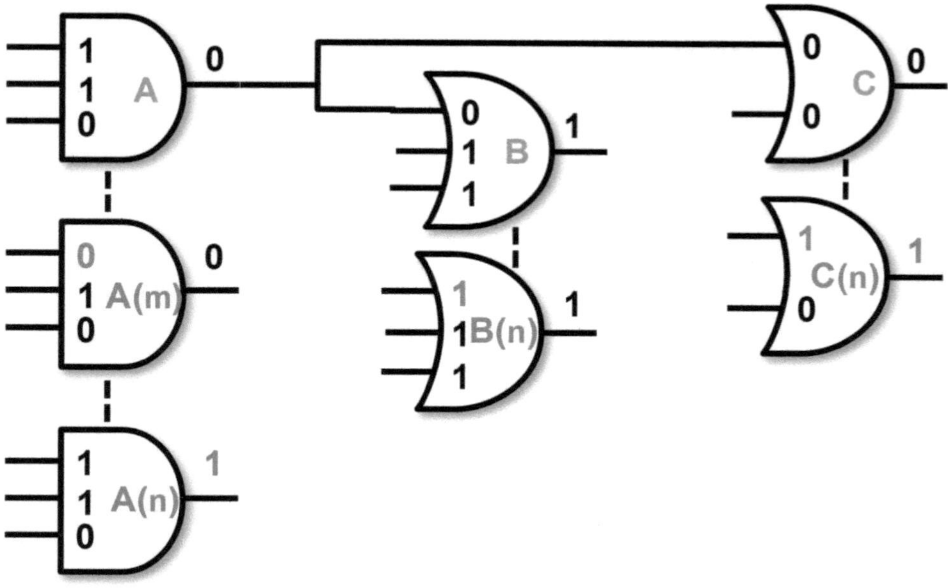

Figure 22: Example for concurrent fault simulation

The faulty machine m represents one input of gate A stuck-at 0. For this a copy of gate A is stored. Since the output value of gate A has not changed compared to the fault-free simulation there is no more activity and the simulation for this fault stops here. A second fault n (output gate A stuck-at 1) is inserted. Since the input values of the succeeding gates B and C have changed compared to the fault-free simulation copies for these gates are stored (B(n) and C(n)). The fault is also active after gate C and the activity cone continues here.

The time and space reduction by the presented measures should be estimated as follows:

1. Equivalence classes: number of faulty machines can be reduced by a factor of 2
2. Fault dropping: it can be assumed that during the simulation as a mean 10% of faults are undetected
3. Event-driven: a mean number of activity cones representing 10% of all gates is gained by experience

Revisiting the formulas for time and space effort:

- Time for simulating a single vector: $T_{fsim} \approx N_{gates} * t + N_{faults} * N_{gates} * t = t * (N_{gates} + 2N_{gates}^2) => O(N^2)$
- Memory for simulating a single vector: $M_{fsim} = N_{gates} * (1+N_{faults}) = N_{gates} * (1+2*N_{gates}) = N_{gates} + 2N_{gates}^2 => O(N^2)$

Replacing N_{gates} and N_{faults} by the estimations for the impact of effort reduction measures results into an overall reduction of ~200. This reduction is valid for both time and space.

The order of complexity is still N^2. Thus, doubling the circuit size results in a run time and memory increase of 4x.

Basic ATPG algorithm and D-calculus

Other than fault simulation ATPG allows the generation of test pattern based on fault models. The basic ATPG algorithm is explained using the two NAND-gate circuit shown in Figure 23. To explain the principle a vector for the stuck-at 0 fault at internal node d is calculated. Nodes a, b and c can be stimulated by ATE and output y is observable at ATE to check for pass or fail of a test.

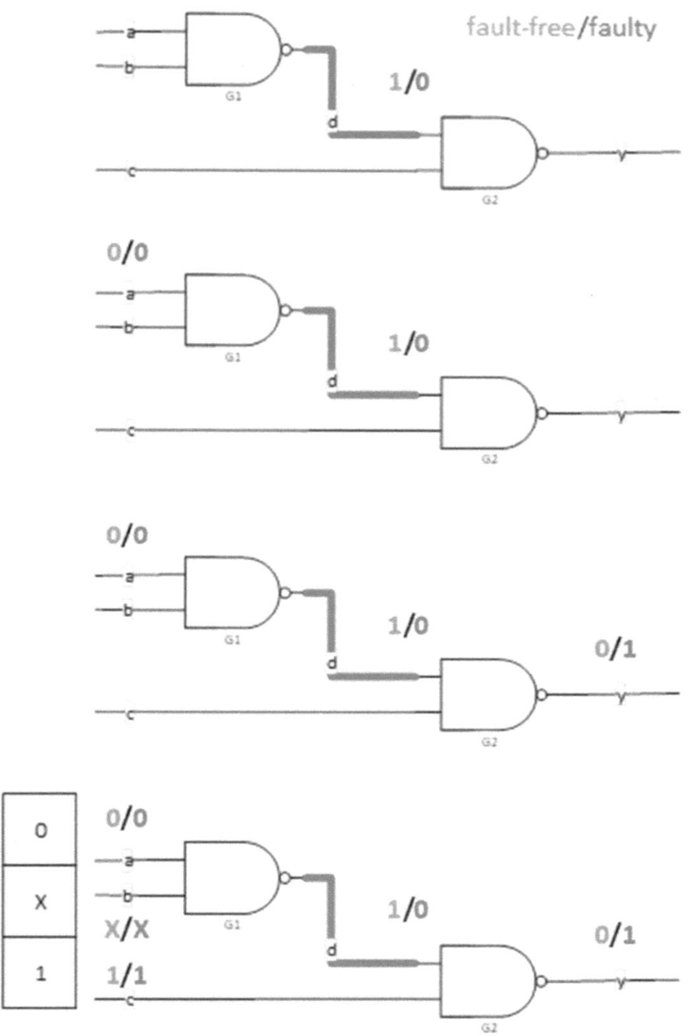

Figure 23: Basic ATPG algorithm for stuck-at 0 fault at line d

Four steps of ATPG are shown in Figure 23. To follow these steps logic tables in the Appendix can be used:

1. Fault initialization: At the fault site (node d) an opposite logic value than the faulty value is required. For a stuck-at 0 a logic 1 is required in the fault-free circuit. This is indicated by the 1/0 notation (fault-free/faulty).

2. Justification: Here the algorithm selects a condition at gate input to get the required value 1 at the output. For the example a=0 is selected. Please note two more options exist. These are b=0 or a=b=0.
3. Fault effect propagation: During this step the fault effect is propagated from the fault site to the output of the circuit. For this small example there is only a single path using gate G2 available for fault effect propagation.
4. Backward implication: This step defines all nodes in such a way that fault effect propagation along the chosen path is possible. For this, node c needs to be set to logic 1 only. It is important that the difference between fault-free and faulty is visible at the output of G2. Choosing a logic 0 for node c would result in y=1 for both fault-free and faulty cases.

The resulting test vector for this specific fault is a=1, c=1 and b is undefined or 'don't care'. In the context of realistic, i.e. larger circuits than the chosen two-gate example, undefined values in test vectors can be used to compress the overall vector set. A typical test vector is able to target more than a single fault resulting in test sets with significant less vectors than faults targeted.

Figure 24 shows the same circuit example with stuck-at 1 fault at fault site d. The four steps to calculate a test vector are as follows:

1. Fault initialization: For fault initialization a fault-free/faulty pair 0/1 is required.
2. Justification: To get d=0 in the fault-free circuit a single input combination a=b=1 is possible only for gate G1.
3. Fault effect propagation: The fault effect is again propagated towards the output y using gate G2.
4. Backward implication: Again c=1 is required to get a difference between fault-free and faulty circuit at the output.

The resulting test vector for this fault is a=b=c=1. Here all input signals are fully specified.

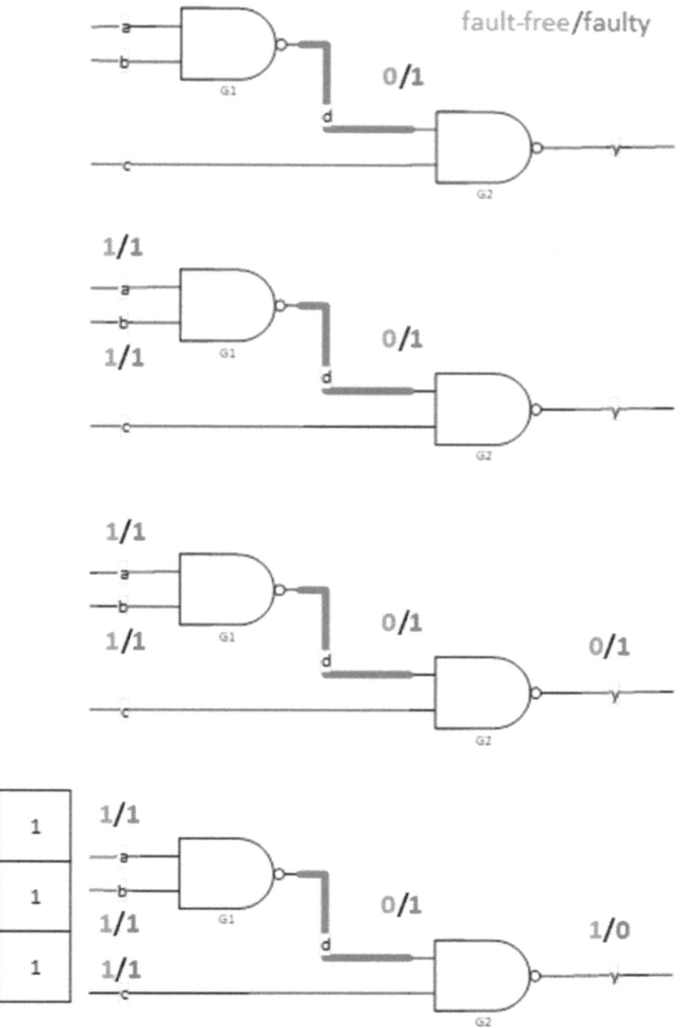

Figure 24: Basic ATPG algorithm for stuck-at 1 fault at line d

The first formalization of the ATPG algorithm was published in [41]. Here Roth introduced 'D' and '$\bar{D}$' as symbols for

- D: Fault-free state 1, faulty state 0 (1/0)
- $\bar{D}$: Fault-free state 0, faulty state 1 (0/1)

For all Boolean gates propagation tables or cubes for D and $\bar{D}$ are defined. The right table within Figure 25 shows such a propagation cube for AND/NAND functions. For other propagation tables see *Appendix*. A D or $\bar{D}$ is propagated if the other input signal has the non-controlling value (=1 for AND/NAND). The controlling value (=0 for AND/NAND) does not propagate a D or $\bar{D}$ to the output. Looking back to the examples above, the fault effect would not propagate if controlling value is applied.

A	B	AND	NAND
0	0	0	1
0	1	0	1
1	0	0	1
1	1	1	0

A	B	AND	NAND
0	D	0	1
D	1	D	$\bar{D}$
$\bar{D}$	0	0	1
1	$\bar{D}$	$\bar{D}$	D

Figure 25: 2-valued logic table and D-propagation table for AND/NAND

A more complex example (Figure 26) to explain the D-algorithm is taken from [42]. The assumed fault is node I8 (output of OR-gate G2) stuck-at 0.

To initialize the fault I4=1 is chosen. The D-values are propagated towards the circuit output via I9 and I10 simultaneously. This propagation is also called 'D-frontier' in the literature. Justifying all signals results in a test cube <a, b, c, d, e, f; w> of <1, X, 1, 1, X, 0; D>. This test cube is combined with test cubes for other faults by the ATPG algorithm to get a compact test set. At ATE 0 and 1 logic values are allowed. The test cubes are converted into test vectors which contain 0 and 1 logic values only. The D and $\bar{D}$ output values are replaced by the fault-free value 1 and 0.

For the example in Figure 26 all decisions made have been proven conflict-

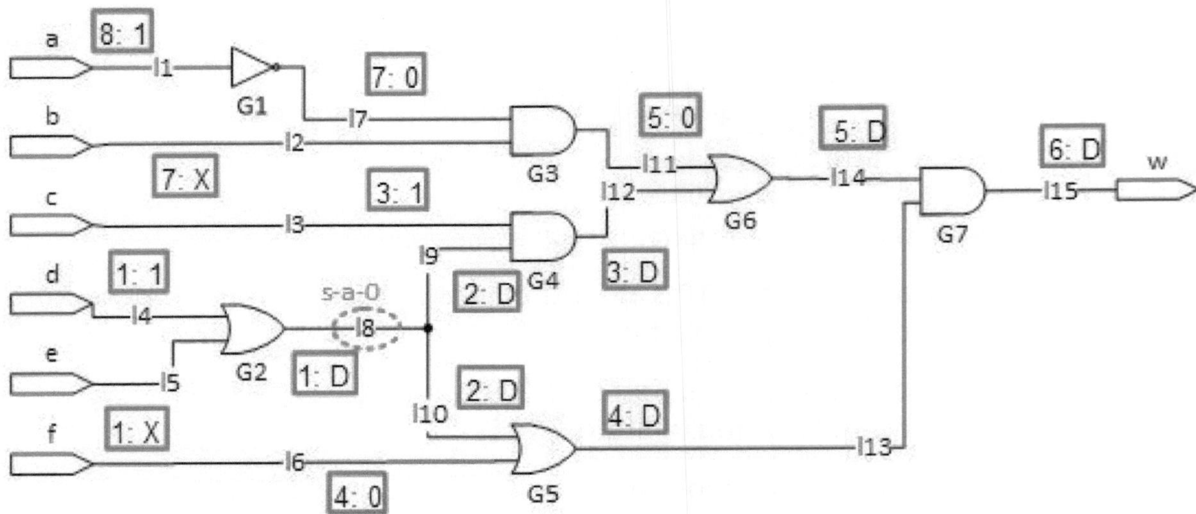

Figure 26: Example for D-algorithm

free. Decisions for this example are:

- Using I4 for fault initialization
- Propagating D via I9 and I10 simultaneously
- Chose I7 to guarantee logic 0 at I11

In case of a conflict a decision has to be replaced by an alternative one. This step is called a backtrack. If all possible decisions are applied without any conflict-free test cube calculation such a fault is called redundant. Using ATPG tools a parameter to limit the number of backtracks is available. By this the runtime effort allowed for a single fault can be limited.

DfT Tool Box

Looking for a definition of DfT results into two main findings. Depending on the view you can see DfT as defined by

- Definition #1: DfT measures spend efforts on Silicon area.
- Definition #2: DfT measures increase the observability and controllability of internal nodes during test.

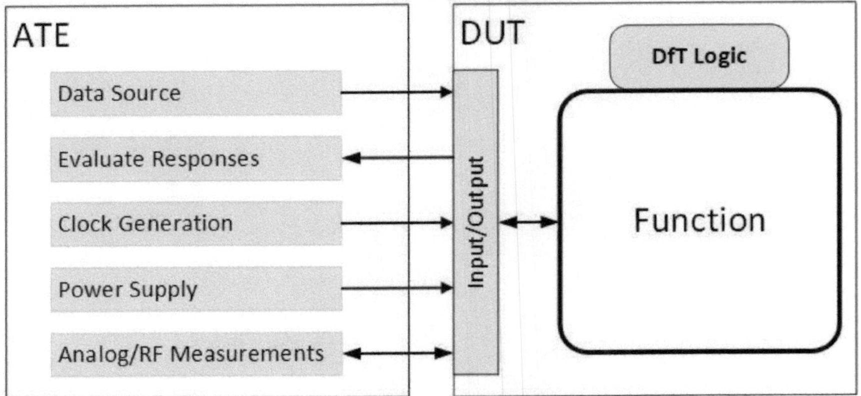

Figure 27: Definition #1 of DfT

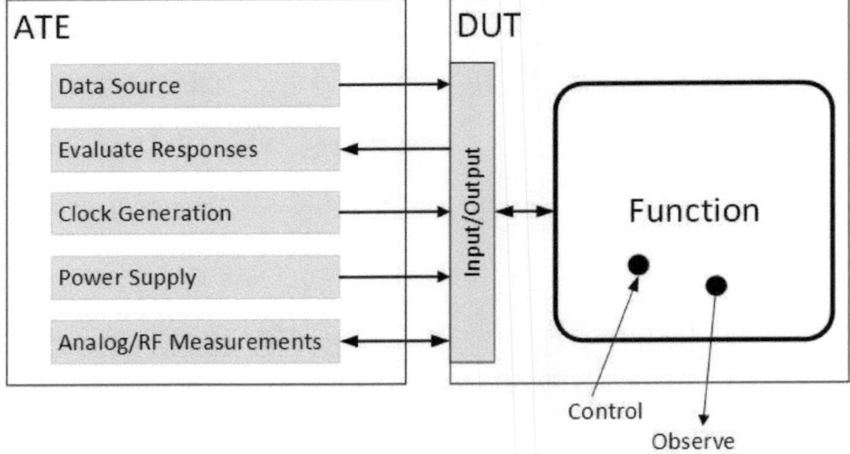

Figure 28: Definition #2 of DfT

There is no contradiction between these two definitions. The effect of DfT is, again independent of the used definition, a reduction of

- Manufacturing test time
- Requirements on ATE features
- Test content development and ATE programming effort

As explained in Chapter Test time, yield, quality and reliability all of the listed benefits result in costs, either as part of CoS or R&D category. Sometimes second order benefits like keeping project schedules or achieving quality goals are influencing decisions towards certain DfT measures. Another aspect is yield which is enabled by redundant structures on chip which are also part of the DfT circuitry. A typical example is the use of memory repair schemes as described in Chapter Memory Test and BIST.

Generally, decisions on DfT should be done from a cost perspective. DfT has to pay-off by test costs and/or test development efforts. Consequently, DfT needs to support manufacturing test. DfT spending for characterization and qualification purposes only has a higher hurdle to be accepted since here only a small number of devices is used.

DfT methods summarized within next chapters are typically specialized for different circuit structures. Viewing the layout picture in Figure 29 embedded memories are most regular structures. Cell arrays of various shapes are generated by compilers. More 'cloudy' structures named µ-Controller and Digital Signal Processing in Figure 29 are standard cell areas representing digital logic generated by synthesis.

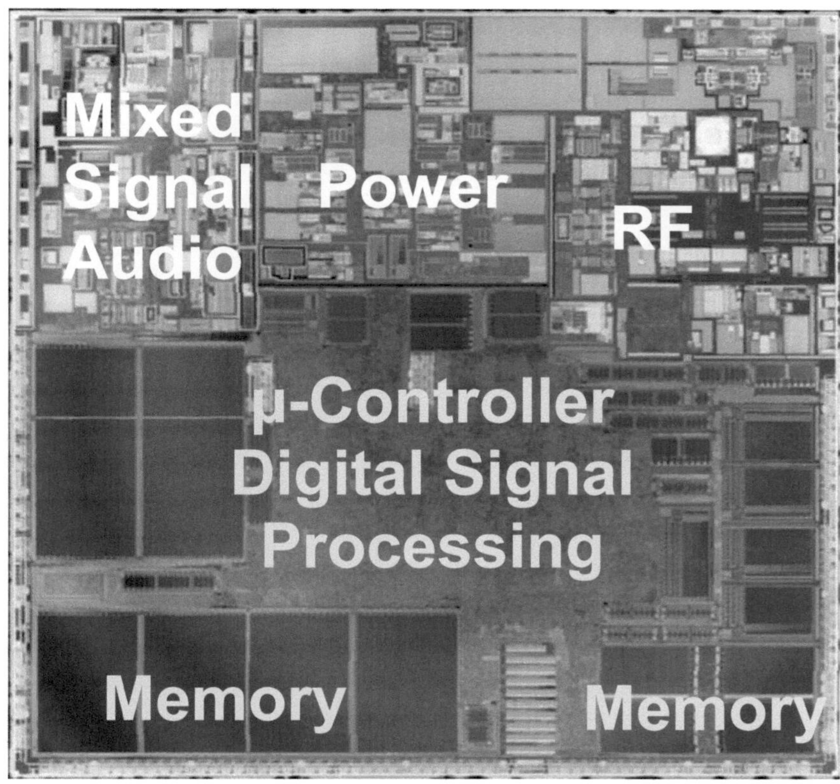

Figure 29: Layout of a System-on-Chip

The top of the layout contains Mixed-Signal Audio, Power and RF functions. Such modules are mainly developed manually as transistor or even layout design. The typical manufacturing test approach for these modules is functional. Test is constructed in such a way that the function and critical parameters are checked for every manufactured device.

For the embedded memory and standard cell-based areas the structural test approach dominates. Structural test does not ask for function: "It's a second level cache memory and I have to test the cache behavior". The structural approach is: "It's a memory, I have to test it using algorithms". Structural test methods are supported by EDA tools and the methods can be used independent of the realized system function.

Scan Design and Automated Test Pattern Generation

Standard cell-based structures are tested by test patterns based on stuck-at and other fault models. These patterns are applied in a so-called scan test mode. Test patterns for scan test are provided by automated test pattern generation (ATPG) software.

The problem description for ATPG is as follows: *Find a compact test set for a given fault model*. ATPG needs a circuit description and a selected fault model as input. The result is a compact test set with a known fault coverage, see Figure 30.

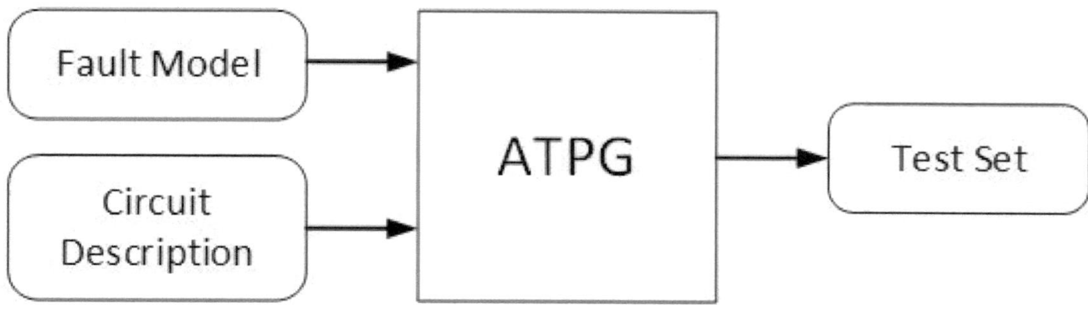

Figure 30: Generalized ATPG problem statement

The most prominent fault model is the single stuck-at fault model. This model considers faults at internal nets of gate level netlists. The presence of a single fault is assumed and ATPG tries to find a test vector for such a fault. A circuit with 1000 nets has roughly 2000 faults. For every net a stuck-at zero (s-a 0) and a stuck-at one (s-a 1) is modeled. Since most of the identified test vectors are able to detect more than a single fault a test set contains less than 2000 vectors for the mentioned 1000-net example. Besides fault coverage the compactness of the generated test set is the second important criterion for the effectiveness of ATPG algorithms and tools.

For complex digital circuits ATPG has been proven effective for combinational logic. For sequential behavior as it is typical for digital logic like finite-state machines (FSM) or processing units the ATPG is more time consuming and high fault coverage is not guaranteed. Complexity theory classifies sequential ATPG as an NP-complete problem [4]. The basic ATPG algorithm is described in Chapter Basic ATPG algorithm and D-calculus. To enable combinational-ATPG procedures for digital sequential circuits scan design was introduced. A dedicated test configuration (scan-mode) allows a serial access to flip-flops (FF) as internal memory elements.

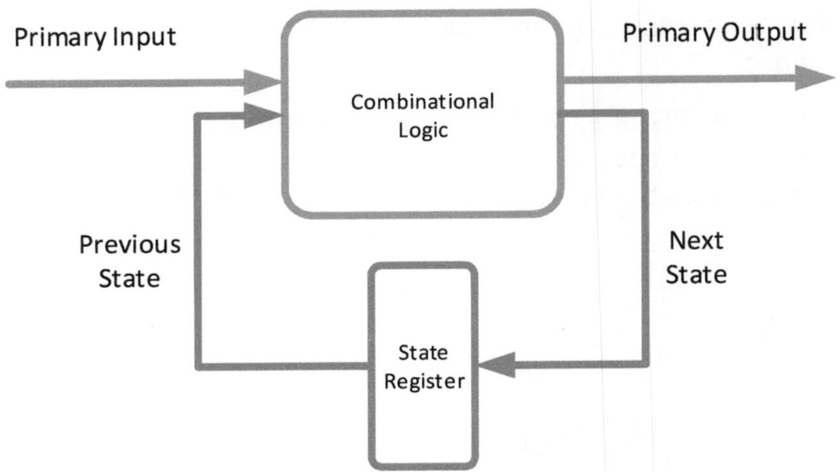

Figure 31: Mealy Automaton

Figure 31 shows a general digital, synchronous sequential circuit. All memory elements are grouped in the state register. This consists out of synchronously clocked flip-flops representing the actual state of the register. The upper part of the circuit contains memory-free combinational logic only.

For scan design all flip-flops inside the state register are replaced by scan- flip-flops (scan FF) as shown in Figure 32. A conventional flip-flop is enhanced by a multiplexer in front of its data input D. The multiplexer is controlled by a scan-enable signal which allows to select between normal function data and scan inputs. To configure a scan chain between two scan flip-flops output Q is connected additionally to the scan input of the next scan flip-flop. Connections of Q outputs to logic gates are left unchanged for normal functional operation. A scan chain configuration is shown in Figure 33.

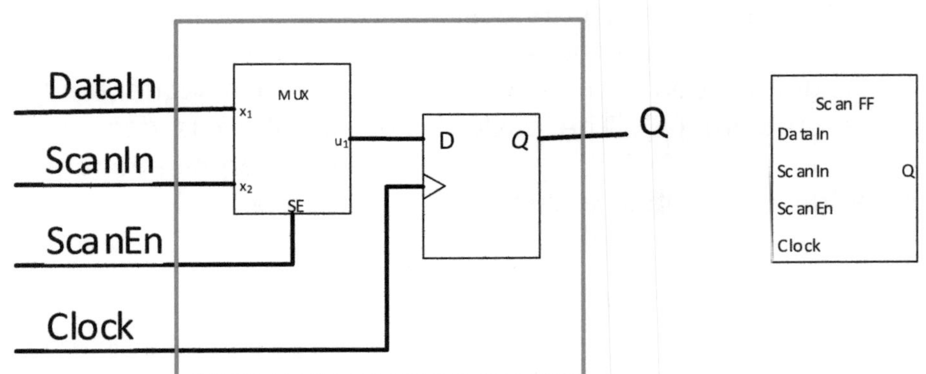

Figure 32: Scan flip-flop of type multiplexed flip-flop

A scan test procedure is distributed into three phases:

1. Serial shift-in of values into the scan chain. This pre-loading of the scan chain defines a known state of the circuit. The outputs Q for all scan flip-flops are defined now as preparation for step 2.
2. Apply data to circuit inputs and capture data in parallel into all scan flip-flops. By this, test results are available at circuit's combinational outputs and in scan flip-flops.
3. Serial shift-out of scan flip-flop data via serial scan chain

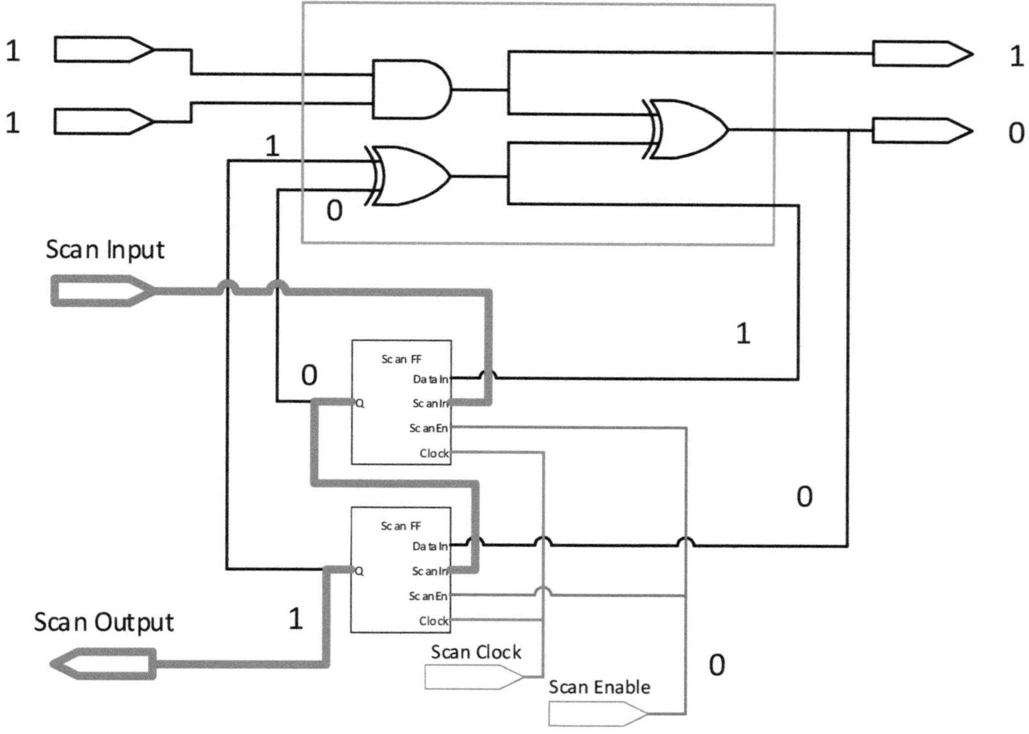

Figure 33: Principle of scan path (capture cycle shown)

The procedure is controlled by a synchronous clock for all cycles and by the scan-enable signal. Scan-enable is active for serial phases 1 and 3. For phase 2 (capture cycle) scan-enable is inactive. This is also the value for normal functional operation.

Data for step 1 (serial shift-in) and for combinational inputs during step 2 are supplied by ATE. Combinational output values after the capture cycle are compared at ATE as well as serially shifted-out data from scan chain.

A test vector generated by ATPG contains values for both data inputs and scan flip-flops applied during the capture cycle. Using scan design the task for ATPG is combinational since no sequential behavior of the circuit needs to be considered.

Since many test vectors are applied reading out the scan chain (phase 3) allows the shift-in of the next vector. Thus, phase 1 of vector n+1 is phase 3 of n.

The number of clock cycles required to apply a single vector is the sum of the number of shift cycles plus a single capture cycle. The number of shift cycles is identical to the length of the scan chains l. Since typically l≪1 the number of cycles for a single vector can be assumed to l. To get all scan test cycles the chain length has to be multiplied by the number of ATPG vectors N:

$$\text{Test cycles} = N*(l+1) \cong N*l$$

To calculate the test time the number of test cycles has to be divided by the used frequency:

$$\text{Test time} = 1/f*N*l$$

The ATE memory required to store the complete scan pattern is identical to the number of test cycles:

$$\text{ATE test data} = N*l \qquad \text{(for a single ATE channel)}$$

It is very important to realize that a capture cycle can be applied only if all chains are loaded completely. Thus, the longest scan chain has to be used for l. The consequence out of this is the requirement to make all chains as balanced as possible. Otherwise test time and required ATE memory are not optimized.

Scan pattern types

The terms ATPG or scan pattern are used synonymously. Such a pattern uses the scan system and ATPG software generates it. Pattern types differ for:

- Fault model used by the ATPG tool
- Dynamic or static type which results in different settings for on-chip clock control during test and for ATE timing
- Special propagation constraints for fault effects, e.g., around embedded SRAMs or along paths

The following table gives an overview about the scan pattern types.

Type	Model	Static/ Dynamic	Comment
Chain shift	none	None	Checks shift capability of chains
Stuck-at	Stuck-at	Static	Basic type without clock-sequential and multi-load
Transition delay	t-delay	Dynamic	Typically separated according to clock domains
Path delay	p-delay	Dynamic	Typically separated according to clock domains
Cell-aware static	Cell-aware	Static	
Cell-aware dynamic	Cell-aware	Dynamic	Typically separated according to clock domains
Timing-aware	Small delay	Dynamic	Typically separated according to clock domains
Clock-sequential	Stuck-at	Both	Add-on to propagate through SRAM
Multi-load	Stuck-at	Both	Add-on to propagate through SRAM
Ram-sequential	Stuck-at	Both	Add-on to propagate through SRAM
IDDQ	Pseudo stuck-at	Static	Scan for stimulation only, observation via current measurement
Bridging	Aggressor-victim	Static or Dynamic	Fault site extracted from layout

Chain shift

The simplest scan pattern is checking the correct shift through the internal scan chains. Typically, a sequence of '0011' is shifted through the scan chains. This pattern is short and has no contribution to any fault model-specific test coverage.

Stuck-at

This is the basic scan pattern type addressing stuck-at faults using the classical scan cycle. A single scan vector is loaded, a single capture clock cycle is applied, and the scan vector is unloaded while the next vector is shifted into the scan chains. This pattern type should be the largest contributor to stuck-at coverage achievement and should have the highest vector count for all stuck-at tests.

Transition delay

Transition delay pattern is generated to detect slow-to-rise and slow-to-fall faults at internal nodes of the circuit. For this a two-capture cycle test is required. The number of fault locations is the same as for stuck-at faults. Since the constraints for fault stimulation and propagation is stricter caused by the two-capture cycle test the resulting test coverage is typically lower than for stuck-at.

To check the dynamic behavior the capture phase has to be executed using system-internal clock speeds. This requires a different configuration of the on-chip clock system and a different ATE timing than applied for static tests like stuck-at. A delay pattern is characteristic for a single clock domain. For multi-clock design separated transition delay patterns are used.

Path delay

A path delay pattern uses the same on-chip and ATE timing settings as used for transition delay pattern. For this pattern type complete paths from scan flip-flop to scan flip-flop are used as fault locations. Since the number of different paths (and by this possible fault locations) is typically orders of magnitude higher than number of internal gates only a sub-set of possible faults is used to generate this pattern.

The same as for other delay pattern types, path delay patterns are separated according to covered clock domains.

Cell-aware static and cell-aware dynamic

Cell-aware pattern types consider cell-internal faults of a standard cell library. For this the layout of a standard cell is investigated and the detection conditions for the assumed internal faults are mapped to cell's inputs and output(s). The library is enhanced by an additional view containing the detection conditions for each internal fault. While some internal faults are detectable by static tests others require dynamic test for detection. For this static and dynamic cell-aware patterns are generated.

Using cell-aware modeling the number of fault locations compared to basic stuck-at and transition delay faults is increased significantly. The modeling accuracy is increased with main benefits for complex cells like ANDOR or ADDER functions.

Timing-aware

ATPG for timing-aware scan patterns uses timing information provided by Standard Delay Format (SDF) files and tries to generate vectors that detect transition faults using the longest detection path. Underlying for pattern generation is the small delay fault model. This model uses slack, which is defined as margin between the delay of a path and the clock period. The slack is the smallest delay fault which can be detected.

The test coverage for this fault model is defined as:

> Small Delay Test Coverage = (Path Delay that ATPG used)/(Longest Path Delay)*100%

The additional consideration of the longest available path results in coverages for the small delay fault model which are lower than for transition faults. There is also significant increase in pattern count and ATPG run time reported. Using launch-off-shift delay test methodology is not supported for timing-aware pattern.

Clock-sequential

This pattern type is used to increase fault coverage of the basic pattern for a specific fault model. Here procedures are applied to detect faults at locations which are not accessible within a single clock cycle. The procedure is as follows:

1. Load scan vector into chains
2. Apply clock sequential cycles (pulse SRAM write/read lines, pulse capture clock of scan chains)
3. Unload scan chains

The number of applied clock sequential cycles is defined by the sequential depth parameter. Typically, this parameter has a single digit value.

Multi-load

To further increase coverage for a new specific fault model a new scan vector is loaded into the scan chains during application of clock sequential cycles. This pattern is an extended version of a clock-sequential pattern as described above.

RAM-sequential

This pattern type uses additional scan loads to write values into two SRAM addresses before executing normal scan cycle. The focus is on fault detection at memory interface signals. If clock-sequential and multi-load pattern are used there is no need to have a RAM-sequential pattern.

IDDQ

IDDQ test uses current measurements to detect circuit-internal faults. The scan system stimulates the fault sites. Fault detection is done by the mentioned current measurements. The pseudo stuck-at fault model is used by the ATPG tool to generate the required pattern. According to this model a fault is detected by IDDQ-test if a fault effect is propagated to the output of a following gate.

Since no further propagation constraints have to be satisfied by the ATPG algorithm the resulting pattern contains typically less than hundred vectors.

Bridging

A bridging scan pattern addresses unintended shorts between interconnect lines within a circuit. Layout extraction is used to limit the number of considered bridges to physically reasonable shorts. A bridge fault between interconnect line A and interconnect line B requires opposite logical values '0' or '1' at these lines. Since physical realities like driving strengths and resistance of a resulting short decide which logical value dominates the aggressor-victim model is used to generate bridging pattern. For this, four vectors are generated for a specific bridge fault:

1. Line A (value '1') is aggressor changing value at victim line B from '0' to '1'
2. Line B (value '1') is aggressor changing value at victim line A from '0' to '1'
3. Same a 1. with aggressor value '0' and victim changing from '1' to '0'
4. Same a 2. with aggressor value '0' and victim changing from '1' to '0'

Victim lines can be checked for static or transition behavior. Thus, both a static or dynamic clocking set ups are used.

Built-In Self-Test

The fully self-testable device is a dream since decades: Just start the device, wait a certain period and get the pass or fail indication. We are used to such kind of self-test for almost all electronic devices. Computers check themselves during the boot period. Various car electronics are tested if engine starts, any fail is indicated by a control lamp. Smartphones, TV sets and other equipment with significant electronic content run some test routines in start-up phases. Some of the concepts summarized here are supporting the mentioned check at system start. The focus of this Chapter is more on built-in self-test (BIST) used during manufacturing test.

Manufacturing test gives feedback to wafer processing and package assembly by details on kind of fails, detailed diagnosis information and concrete parameter values for voltage, current, timing and others. For system self-test such detail level is out of scope. Typically, achievable coverages are lower in system test and execution is longer than on throughput oriented ATE. Implementing BIST for system test or even as customer feature has additional issues to be addressed, like control by firmware or customer software.

For manufacturing test on-chip BIST circuitry can help to optimize the overall test cost. In Figure 34 the ATE-DUT interaction model is modified. Some data source elements are moved from ATE into the circuit. Also, some evaluation tasks are done on-chip. There are also BIST methods for analog/RF test available. The ATE is still needed for test not executed via BIST logic as well as for clock and power provision.

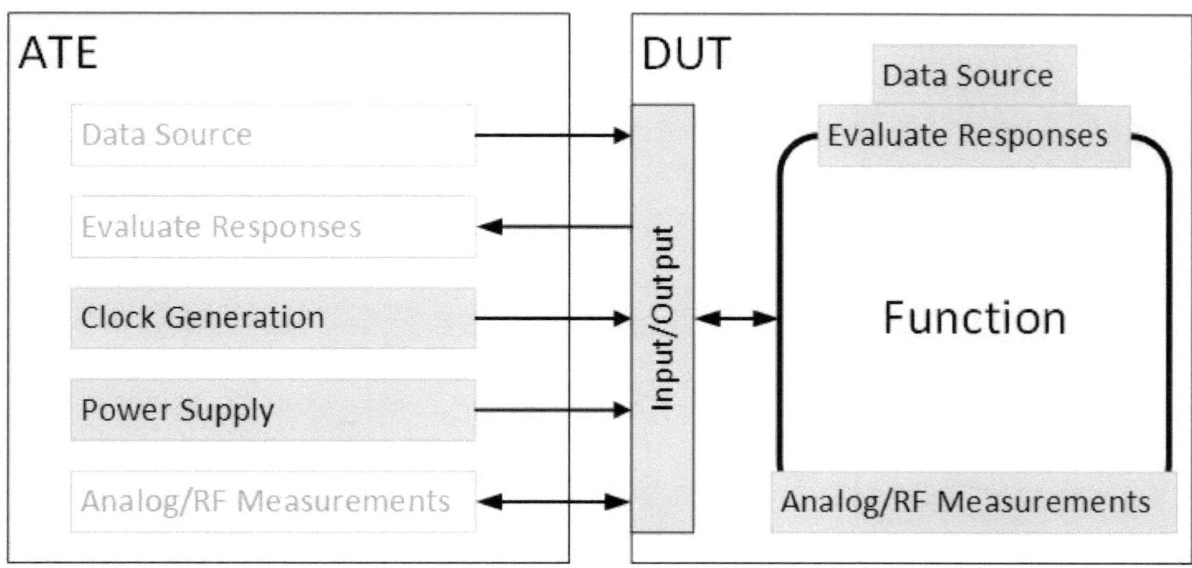

Figure 34: BIST principle

Figure 34 shows some functions of the ATE moved into the functions of the circuit-under-test. This results into less expensive ATE. Revisiting the test cost formula introduced in Chapter Test time, yield, quality and reliability:

Test cost = ATE Cost Rate * 1/Throughput

BIST leads to a reduction of the factor for ATE Cost Rate. The increase of Throughput (or reduction of test time) is enabled by parallel execution of test enabled by BIST logic. Area for BIST logic increases the silicon real estate requirements.

Logic BIST

Logic or standard cell-based structures are usually tested via scan test. To implement BIST for these areas on-chip circuitry to load scan chains and to analyze scan chains' content after capture cycle is required. In Figure 35 scan chains are stimulated by a Test Data Generator module. The outputs of scan chains are connected to a Test Response Analyzer module. The concept of Figure 35 was introduced in [5] as STUMPS (**S**elf-**T**est **U**sing a **M**ISR and a **P**arallel Shift Register **S**equence Generator).

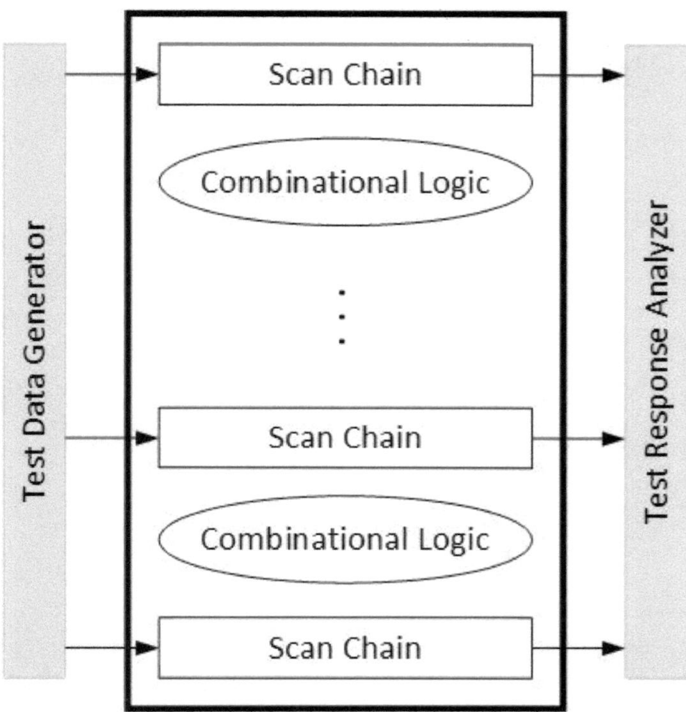

Figure 35: Combining scan test and BIST

Test data generation

The core of the Test Data Generator is a Linear-Feedback Shift Register (LFSR) providing pseudo-random patterns. To explain the principle a 5-bit LFSR is shown in Figure 37. Shift direction for the registers is from left to right. EXOR circuitry is used inside the feedback network.

A 5-bit shift register is able to represent $2^{register\_length} = 32$ different states. If the register starts with the all-zero state clocking the register would result again in the all-zero state. Starting with at least a single bit in one state as

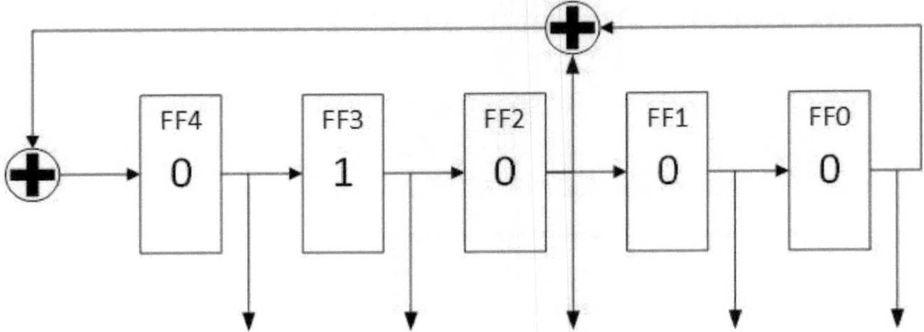

Figure 37: Linear-Feedback Shift Register (LFSR)

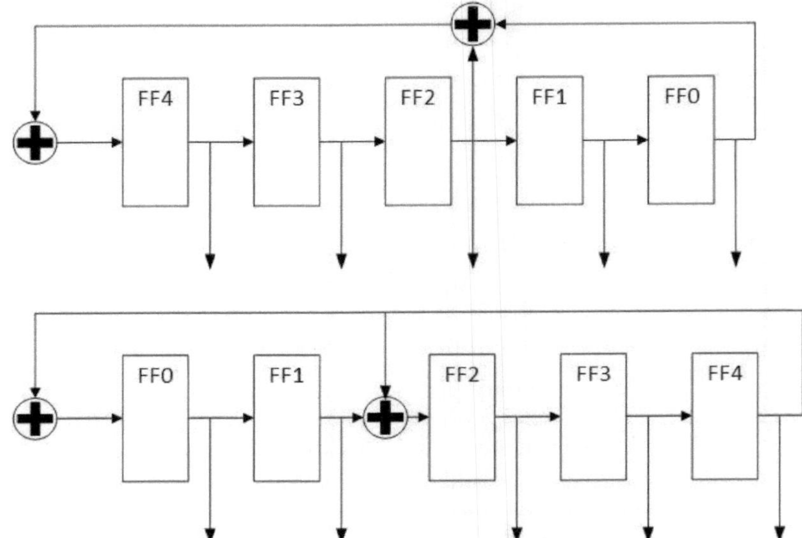

Figure 36: External and internal feedback realization for a primitive polynomial

shown in Figure 37 let the LFSR run through multiple states. The all-zero state is not included within this sequence. The maximum sequence length is $2^{register\_length} - 1$. As long as a clock signal is applied the sequence is repeated. The maximum sequence length is generated only if the feedback network represents a primitive polynomial. The primitive polynomial gives the EXOR positions within the feedback network and can be found for various register length in tables, e.g., in [6].

Two realizations for the primitive polynomial $x^5 + x^2 + 1$ are shown in Figure 36. Both realizations share the behavior to generate sequences of maximum length $2^{register\_length} - 1 = 31$. Nevertheless, the order within the sequences is different.

During scan test with ATPG pattern applied by external test equipment succeeding vectors are independent of each other. A pseudo-random pattern generator (PRPG) has correlations of vectors within the generated sequences.

...	...	...
0001100000110111	**0101010111010001**	**0000000001011011**
0000110000011011	**1101010000011011**	**0000000001011100**
0000011000001101	**0001011000101010**	**0000000001011101**
1000001100000110	**0011001101000001**	**0000000001011110**
0100000110000011	**0110111001100011**	**0000000001011111**
0010000011000001	**1010111110110110**	**0000000001100000**
0001000001100000	**1000110110100011**	**0000000001100001**
0000100000110000	**1101100010010110**	**0000000001100010**
0000010000011000	**0000110101110011**	**0000000001100011**
...	...	...
LFSR	CA	Counter

Figure 38: Sequences generated by different on-chip generators

For 16-bit registers cutouts of sequences are shown. On the left an LFSR sequence with diagonal walking '1s' is given. The sequence in the middle indicates a better 'randomization' than the LFSR sequence with local spots of '1s' and '0s'. Such sequences are generated by cellular automata as presented in [7]. A bad example for a random sequence is shown on the right side which is generated by a counter. Here the least-significant bit (LSB) is changing its

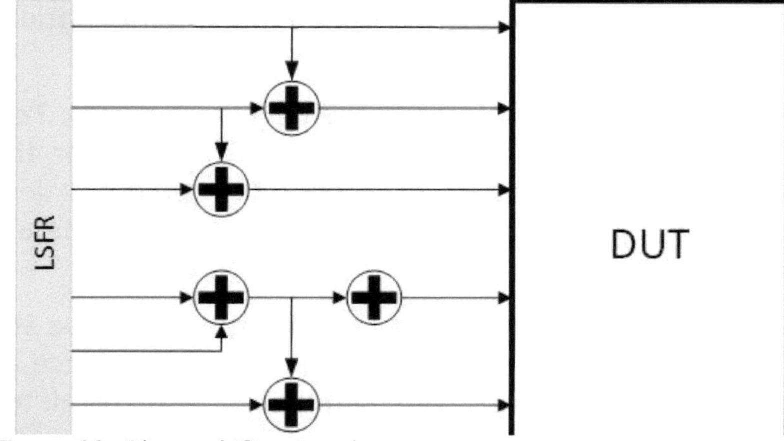

Figure 39: Phase shift network

value with every cycle. The most-significant bit (MSB) would change its value only once during the whole sequence. To summarize, PRPG have some limitations since linear dependencies exist between generated vectors. Effective vectors for fault detection may be far from each other within a sequence. Typically, only a subset of the sequence can be applied. To overcome these limitations some solutions have been proposed.

Phase shift networks between LFSR and scan chains are used to mitigate linear dependencies of LFSR sequences. Figure 39 shows an example. Phase shift networks can be constructed in such a way that an equal probability for '1' and '0' is available on scan chain inputs [8].

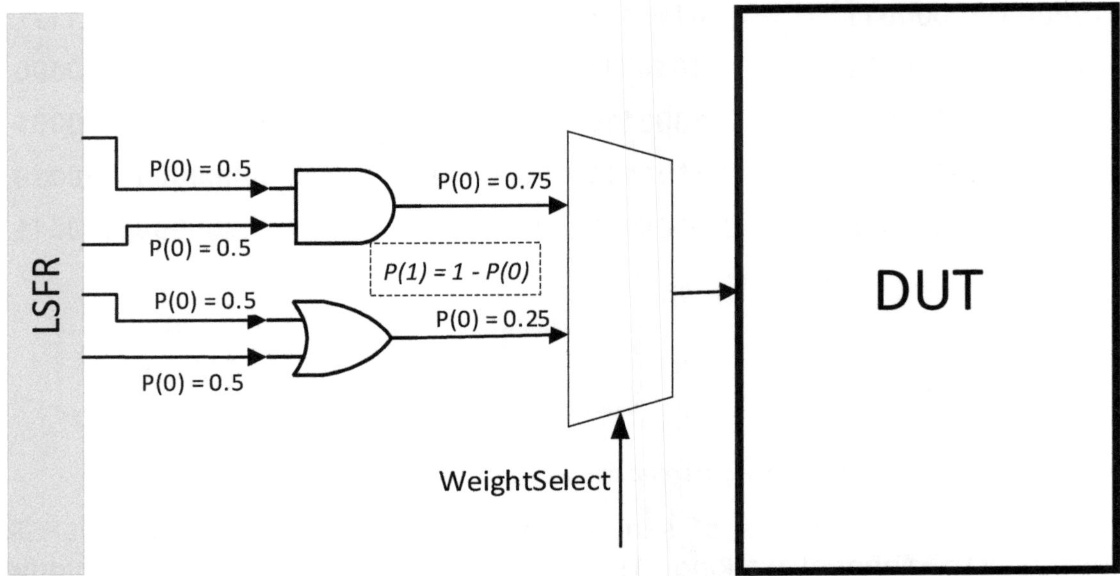

Figure 40: Weighted pattern generation

The principle for weighted pattern generation is shown in Figure 40. This concept allows to skew the distribution of '1s' and '0s' and the inputs of scan chains or other circuit inputs. Weights and weight selection requires simulation-based analysis, and the implementation is circuit-dependent.

Modifications inside circuits-under-test are implemented using test points. Test points as shown in Figure 41 can either enhance the controllability or the observability of an internal node. The impact of observation points is less intrusive than for controllability points. For controllability additional gates are required within the functional path, thus impact on timing is higher.

A more recent publication [9] shows the impact of test points to increase the fault coverage or reduce test time especially for Logic BIST schemes. Since all on-chip pattern generators used for LBIST provide sequences with correlated vectors, achievable fault coverages are typically lower than using ATPG vectors supplied by external ATE.

Signature analysis

The Test Response Analyzer in Figure 35 is typically realized as Multiple-Input Shift Register (MISR). The STUMPS concept presented there has the MISR in its name. The principle of signature compression realized by MISR circuits is shown in Figure 42. As already explained for LFSR a primitive polynomial is also used for the feedback network of a MISR. A sequence provided by the

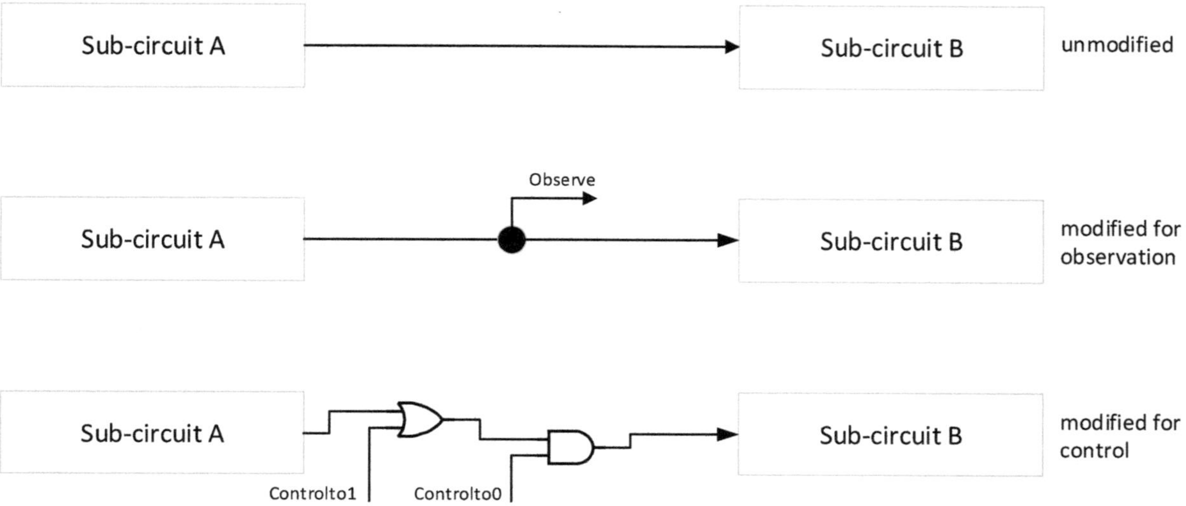

Figure 41: Principle of observe and controllability test points

outputs of a DUT is 'added' to the register content via EXOR gates between the register stages. A typical sequence consists out of thousands of cycles. At the end of the sequence, a fixed-length signature represents the sequence. The signature after test can be compared with a stored correct sequence and used for pass or fail decision.

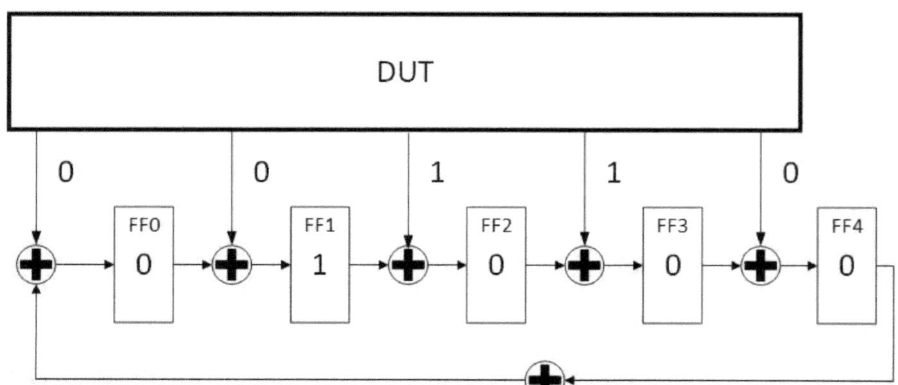

Figure 42: 5 bit Multiple-Input Shift Register (MISR)

Signature compression is a non-reversible process. To evaluate the risk to escape any fault contained in the sequence is explained with the help of Figure 43.

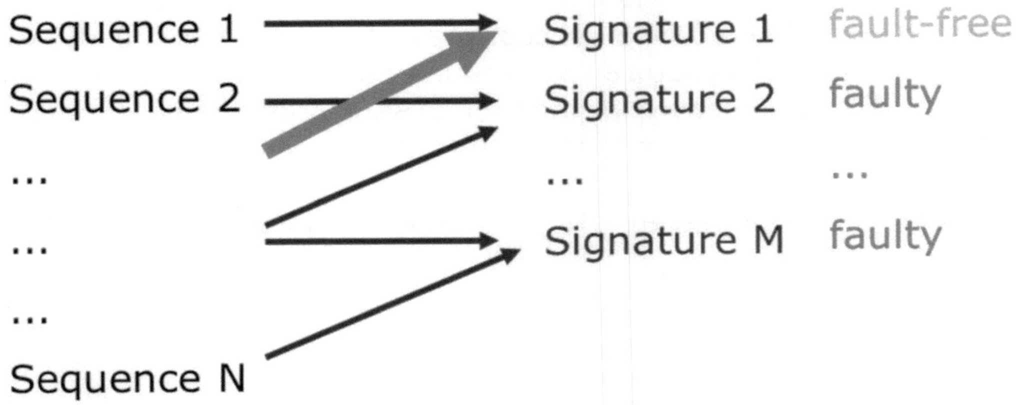

Figure 43: Principle of signature compression

The number of different signatures is limited to $2^{register\_length} = 32$ different states for the 5-bit example. The number of different sequences for a test of many thousand cycles is significantly higher, thus N»M.

The number of possible sequences N is given by the length k of a data stream which is given to the MISR inputs. For k cycles 2^k different combinations for sequences are possible. To get the probability that a faulty sequence is compressed into the fault-free signature (probability for the bold arrow in Figure 43) following terms are used:

- number of possible faulty sequences: (2^k-1)
- number of sequences/signature: $2^k/2^r = 2^{k-r}$
- $(2^{k-r}-1)$ faulty sequences mapped to the fault-free signature, (-1) for fault-free sequence

With these terms the so-called aliasing probability can be formulated as:

$$Paliasing = \frac{2^{k-r} - 1}{2^k - 1} \approx 2^{-r} \text{ for } (k \gg r)$$

The length r of the MISR is the dominating parameter assuming long test sequences. Using a register length of r=20 bit the quality applying a MISR test compared to cycle-by-cycle check during test is 1 dppm ($2^{-20} \approx 10^{-6}$ => 1 dppm test quality reduction). Typical implementations use a MISR length of 32 bit and by this the quality impact is orders of magnitude lower.

Another important implementation aspect is the fact that MISR are unable to recover from any 'X'- or unknown-value capturing. Sources of unknowns for simulation are e.g.:

- Multiple drivers for a bus
- Non-scan state elements
- Non-fully decoded RAM's
- Black box models

The propagation of unknown values into the MISR has to be prevented. For this the circuit is modified by X-bounding logic.

Scan compression techniques

Scan compression techniques have been developed to address the ever-increasing volume of scan test data. This volume increase is driven by more complex digital logic and by demand for increased pattern quality, i.e., higher coverages and more fault models need to be addressed. For scan test the number of flip-flops is the indicator for the logic's complexity. Higher coverages and additional fault models result into higher vector count for the scan patterns. The classical way is to compensate the higher scan data volume by additional vector memory at the ATE and by more test cycles resulting in higher scan test time. Both, additional vector memory and higher test time, lead to in higher test costs.

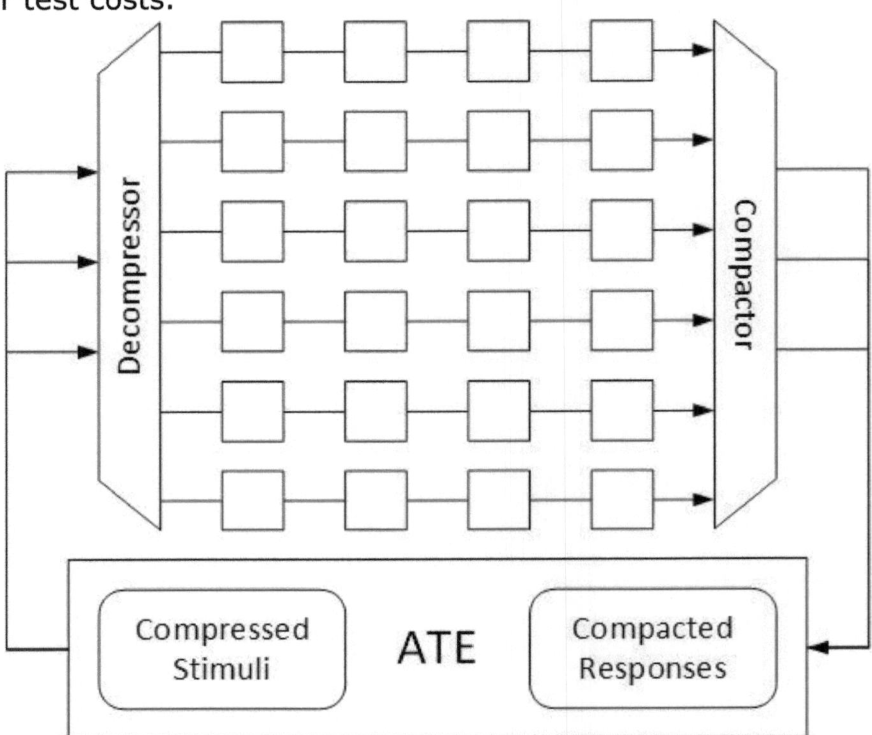

Figure 44: Basic scan compression concept

The basic concept of scan compression is shown in Figure 44. Additional on-chip logic blocks, decompressor and compactor, are implemented to allow the storage of compressed stimuli and compacted test responses on ATE. Driven by the basic scan requirement of short scan chains, the idea is to load more internal scan chains than ATE channels are used or available. The approach is successful if the added on-chip logic is less costly than a more expensive ATE and a higher test time.

Decompression

The first decompression scheme presented was named Illinois scan or broadcast decompressor [10]. The idea is shown in Figure 45 for two ATE channels and 18 internal scan chains. Using these numbers, a compression rate can be defined as ratio between on-chip scan chains and used ATE channels. For this example:

$$r = \#internal\_chains/\#external\_channels = 18/2 = 9$$

Without the simple broadcast decompressor only 2 scan chains could be formed with 9x the length compared to the chain length of Figure 45. The requirement of short scan chains is realized.

Using compression ratio r the formulas for scan test cycles, test time and ATE test data volume can be modified as follows:

$$\text{Test cycles} = N_1 * l_0/\mathbf{r}$$
$$\text{Test time} = 1/f * N_1 * l_0/\mathbf{r}$$
$$\text{ATE test data} = N_1 * l_0/\mathbf{r}$$

Here l_0 is the length of scan chains without any compression. N_1 represents the number of ATPG vectors generated for a specific coverage target. The ATPG algorithm has to consider the specific constraints given by the decompression scheme. It is expected for this simple scheme that the number of vectors N_1 is significantly higher than for a classical scan design. In Figure 45 the upper 9 chains have identical values, also the lower group of 9 scan chains. These are additional constraints for the ATPG process given by the compression configuration. Besides the higher number of vectors N_1 there is a high probability that not all faults are detectable using this decompression logic.

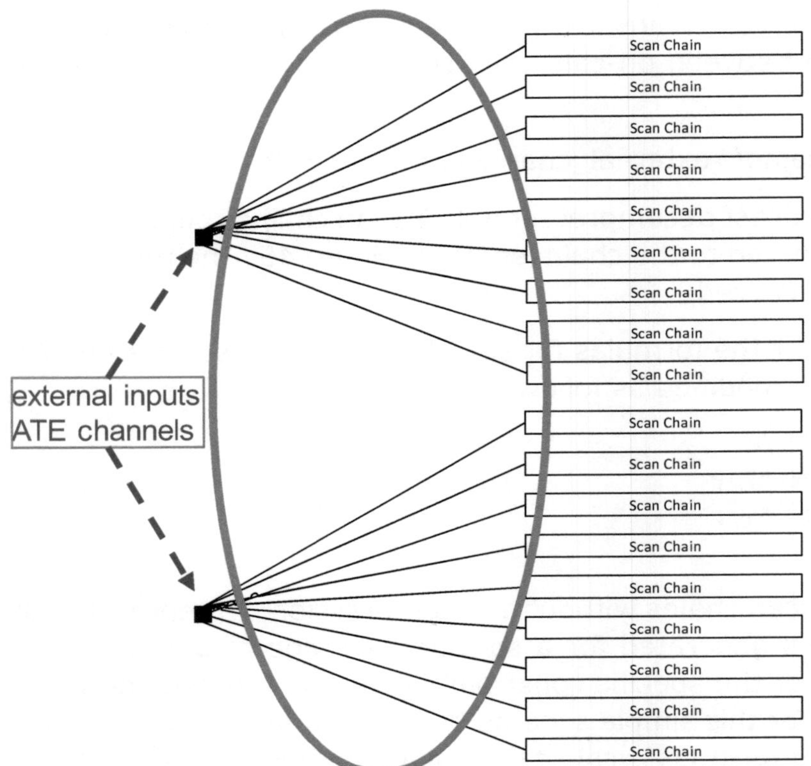

Figure 45: Broadcast decompressor (Illinois scan)

To overcome the problem of identical content for a group of scan chains a more elaborated decompressor was proposed in [11]. The compression technology is named Embedded Deterministic Test (EDT) and the decompressor consists out of two building blocks as shown in Figure 46. The ring generator has a fixed size, and the phase shift network adapts to the number of internal scan chains.

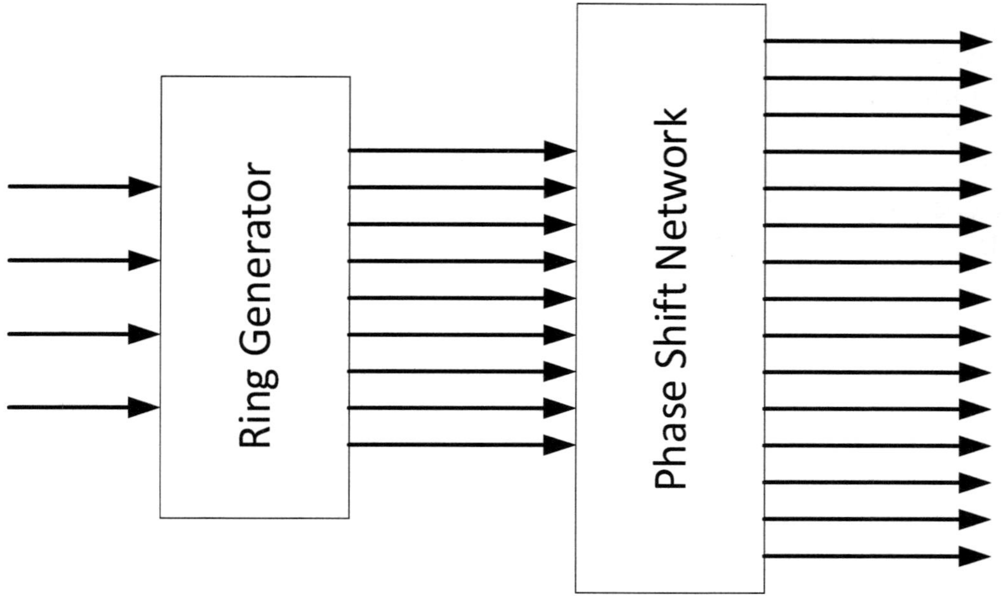

Figure 46: EDT decompressor elements

The realization of the ring generator is shown in Figure 47. There are 32 flip-flops connected serially in an LFSR type manner. Data from ATE are fed into the ring generators via EXOR gates. The outputs of all flip-flops are connected to the inputs of the phase shift network.

An example for a phase shift network is shown in Figure 39. The network consists out of EXOR gates only. Since the ring oscillators have a pseudo-random generator behavior the phase shift network helps to reduce the correlation of scan chain content.

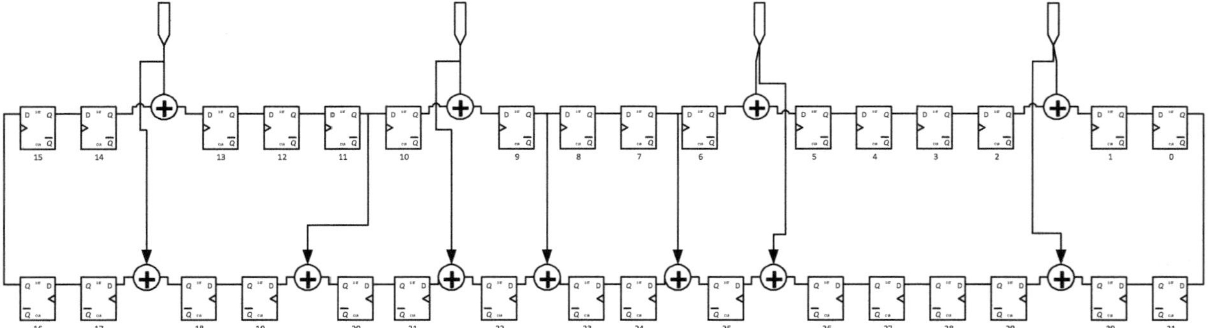

Figure 47: EDT ring generator

The ATPG algorithm has to consider the structure of the decompressor logic. The principle is explained using the 8-bit ring generator example shown in Figure 48, also taken from [11]. Flip-flops and EXOR gates are building up the ring generator logic. These are connected to the phase shift network. Inputs from ATE are input to the ring generator. The outputs of the phase shift network are connected to the inputs of the scan chains SC1, SC2, SC3 and

SC4. The associated bit streams are indicated by variables a_t and b_t. For each input of a scan chain there is a linear equation describing the dependencies from ATE inputs a and b for different cycles. A term like $...b_i \oplus a_k \oplus a_l \oplus b_m$ shown in Figure 48 is constructed for all scan chain inputs. If the ATPG algorithm identifies a test cube for an assumed fault, a set of linear equations available for all scan chains is used to calculate the required ATE data to detect that specific fault.

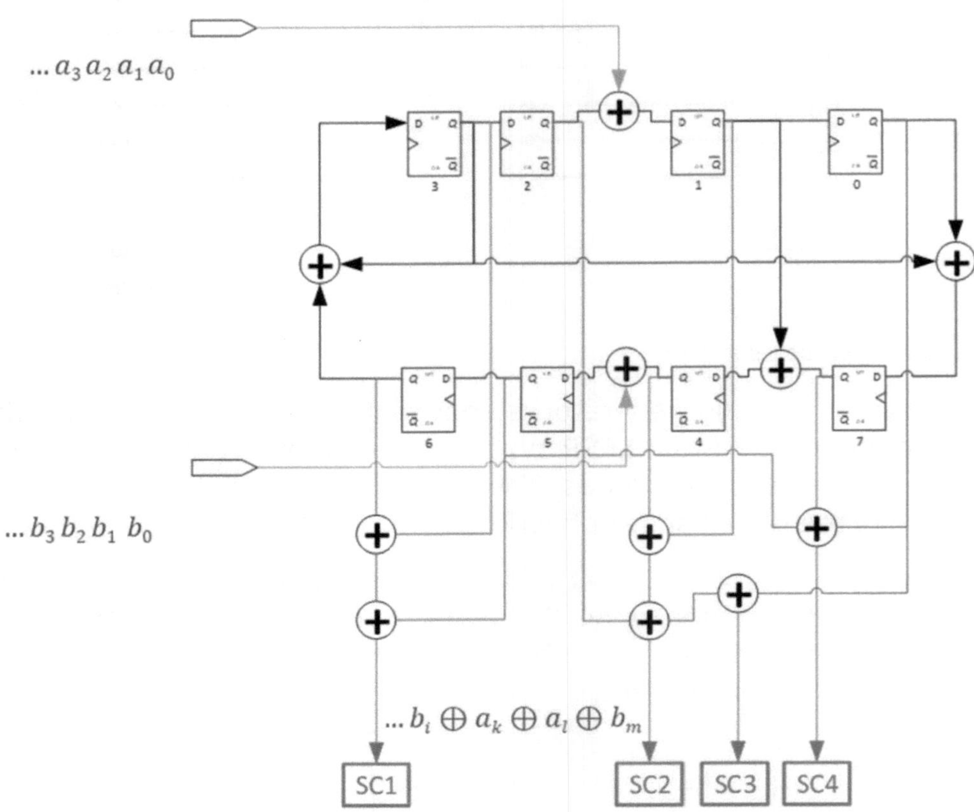

Figure 48: 8-bit ring generator with linear equations

The underlying concept of compression is based on the observation that the typical fill rate for of test cubes for a single fault is within the range of 1-2%. This is shown in Figure 49. A fault within the combinational logic (marked by the asterisk) is stimulated by the upper scan chain requiring two defined bit positions only. The lower scan chain captures the D front at again two bit positions. Test cubes for multiple faults can be compressed together.

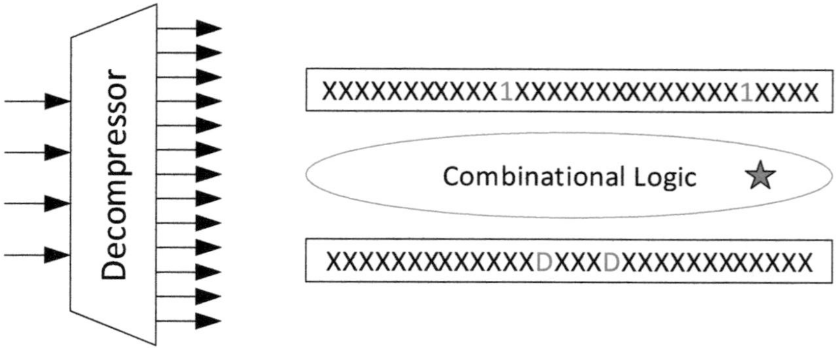

Figure 49: Typical fill rate of a test cube

The fill rate gives also the maximum for the compression rate. If we assume 1% fill rate the compression ratio, i.e., the ratio between internal scan chains and external ATE channels is ~ 1/1% = 100. Nevertheless, this maximum compression ratio is circuit dependent. Experiments show that increasing the compression ratio significantly higher than 100 would result in an inflated ATPG vector count. Since the overall target is to minimize the number of test cycles (= N*l/r) the increased r is over-compensated by an even higher ATPG vector count N.

Compaction

To complete the idea of scan compression according Figure 44 an on-chip compactor has to be implemented. Generally, a MISR can be used to compact test data on-chip. Since this would reduce the test resolution compared to a cycle-by-cycle comparison on ATE an EXOR-tree is typically used. Diagnosis of failing cycles and failing scan chains is also easier using an EXOR-tree. Figure 50 shows an EXOR-tree with 8 inputs.

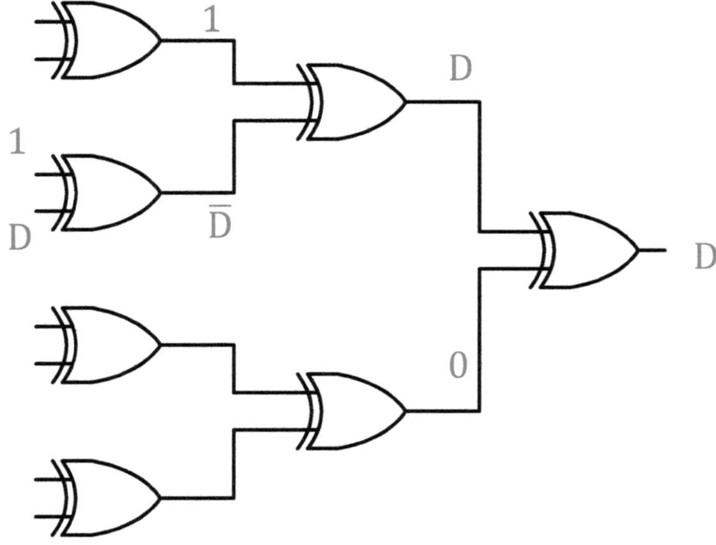

Figure 50: EXOR-tree

Also shown in Figure 50 is the high transparency for D-values which are required as fault detection condition. The output of the EXOR-tree is directly evaluated on ATE. The downside of an EXOR-tree is also a high transparency of X-values. If one of the inputs receive an X-value it will be propagated to the output with absolute probability. Therefore, intrusion of X-values into the EXOR-tree must be avoided.

A concept for compactor logic with X-bounding logic is shown in Figure 51. The ATPG process has the information about X- (or unknown-)values for all scan chains and knows the exact cycles. A mask pattern using this information is provided with the scan test pattern and the implemented logic suppresses the output of a scan chain for all cycles with unknown values. This prevents the capturing of X-values into the EXOR-tree.

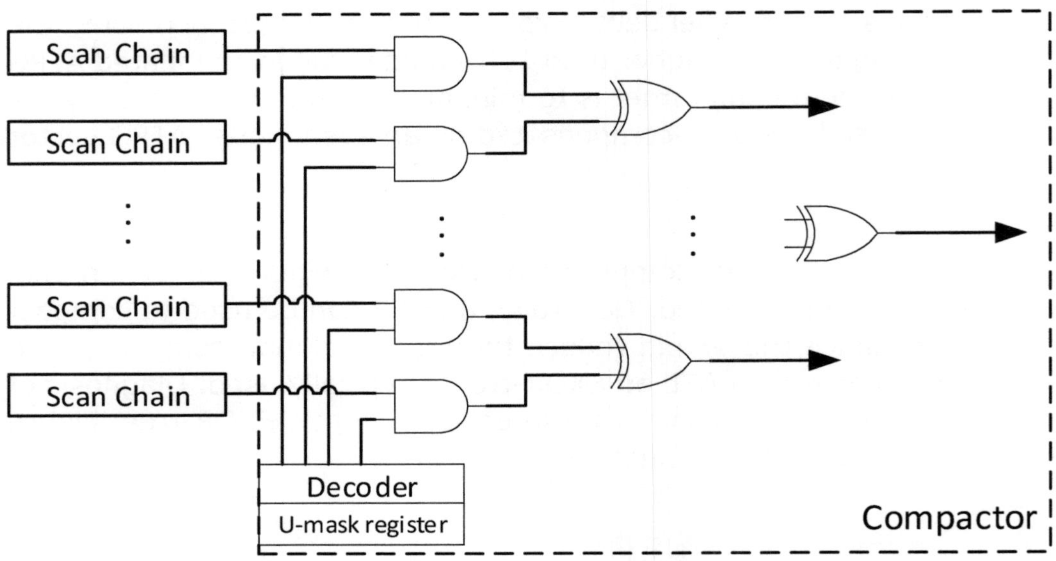

Figure 51: Compactor with X-bounding logic

Modular scan and scan compression units

In a typical System-on-Chip (SoC) for each hierarchical block a dedicated compression system using a decompressor and compactor according to Figure 52 is implemented. This reduces demand for top level routing of internal chains and allows basic testability checks on block level. Modular scan and compression are a prerequisite for hierarchical ATPG introduced below. Figure 52 shows a SoC with two hierarchical blocks including a dedicated compression system for each block and one compression module for all top-level scan chains.

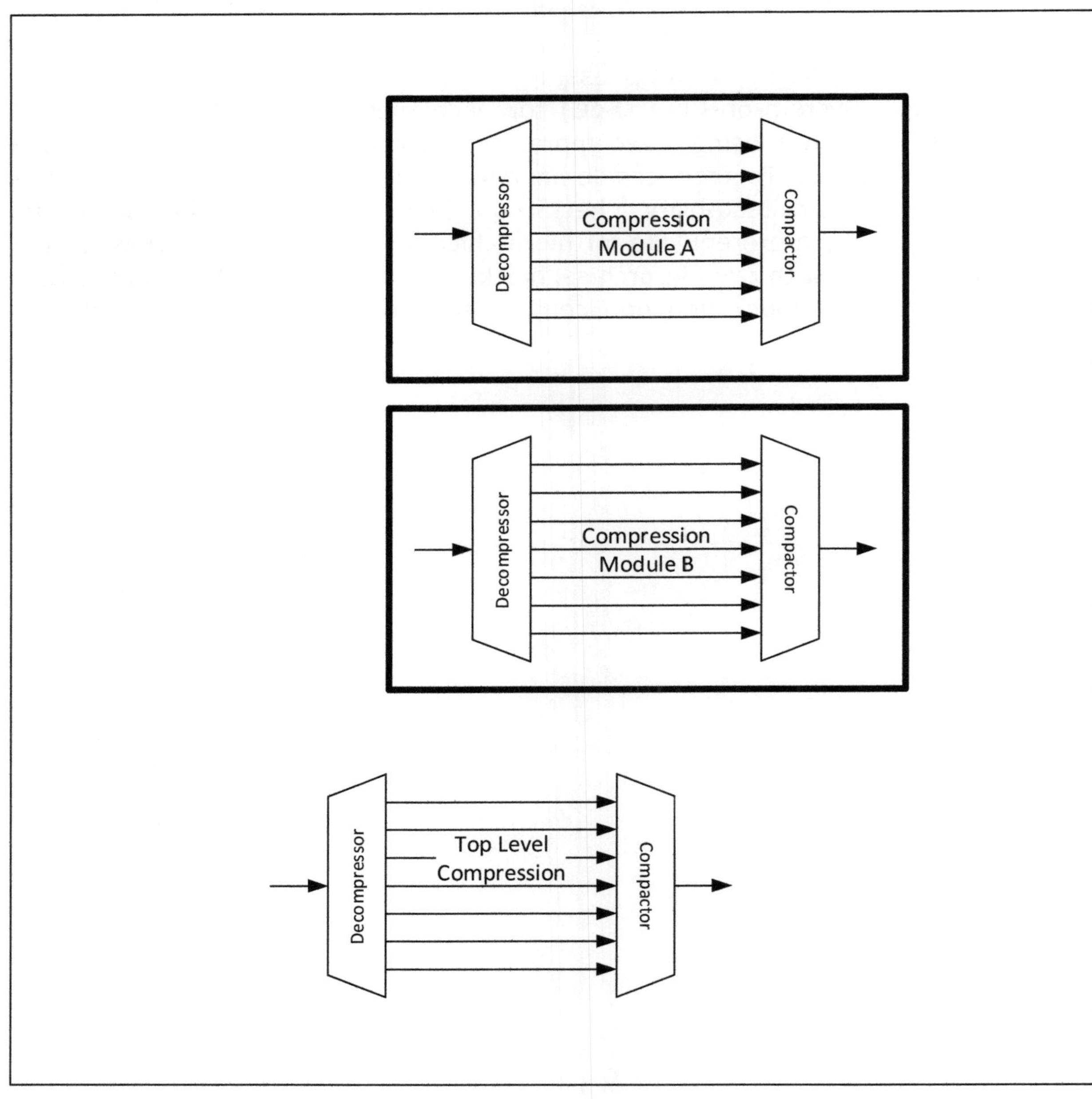

Figure 52: Modular compression system

Hierarchical ATPG

The Classical ATPG flow is as follows:

1. Setup ATPG on module level, check scan design rules like clock, reset and chain tracing
2. Run ATPG on module to identify fault coverage and pattern count for this module, give feedback to RTL designer if fault coverage is too low and/or vector count too high
3. Redo ATPG steps using chip level netlist with all modules

4. Check fault coverage and vector count for complete design, initiate design changes, re-synthesis or ECO implementations
5. Rerun ATPG until fault coverage and vector count is within limits
6. Verify generated ATPG pattern with full timing by gate level simulation, consider all relevant sign-off corners

Redo ATPG on chip level is required to address logic between modules. Flip-flops inside a module are connected to logic outside the module. These connections are needed to control and observe the logic outside the respective module.

Hierarchical Scan (or hierarchical ATPG) is an approach to avoid pattern generation and re-simulation runs on the complete netlist. This significantly reduces the elapsed time for final pattern generation. A flow for hierarchical ATPG is as follows:

1. Setup ATPG on module level, check scan design rules like clock, reset and chain tracing
2. Run ATPG on module to identify fault coverage and vector count for this module, give feedback to RTL designer if fault coverage is too low and/or vector count too high. This mode is called INTEST.
3. Not needed:
    - Redo ATPG steps using chip level netlist with all modules
    - Check fault coverage and vector count for complete design, initiate design changes, re-synthesis or ECO implementations
    - Rerun ATPG until fault coverage and vector count is within limits
4. Verify generated ATPG pattern for module with full timing by gate level simulation, consider all relevant sign-off corners (INTEST runs)
5. Setup ATPG for complete chip level by using Gray-Box models for modules addressed by INTEST runs above and target logic on chip level only. This mode is called EXTEST.
6. Retarget patterns generated by INTEST runs

To allow encapsulation of modules for INTEST runs additional logic is required. The bold indicated scan chains in Figure 53 are used for both INTEST and EXTEST runs. To enable this some mode switching circuitry is required. The bold chains are called wrapper chains.

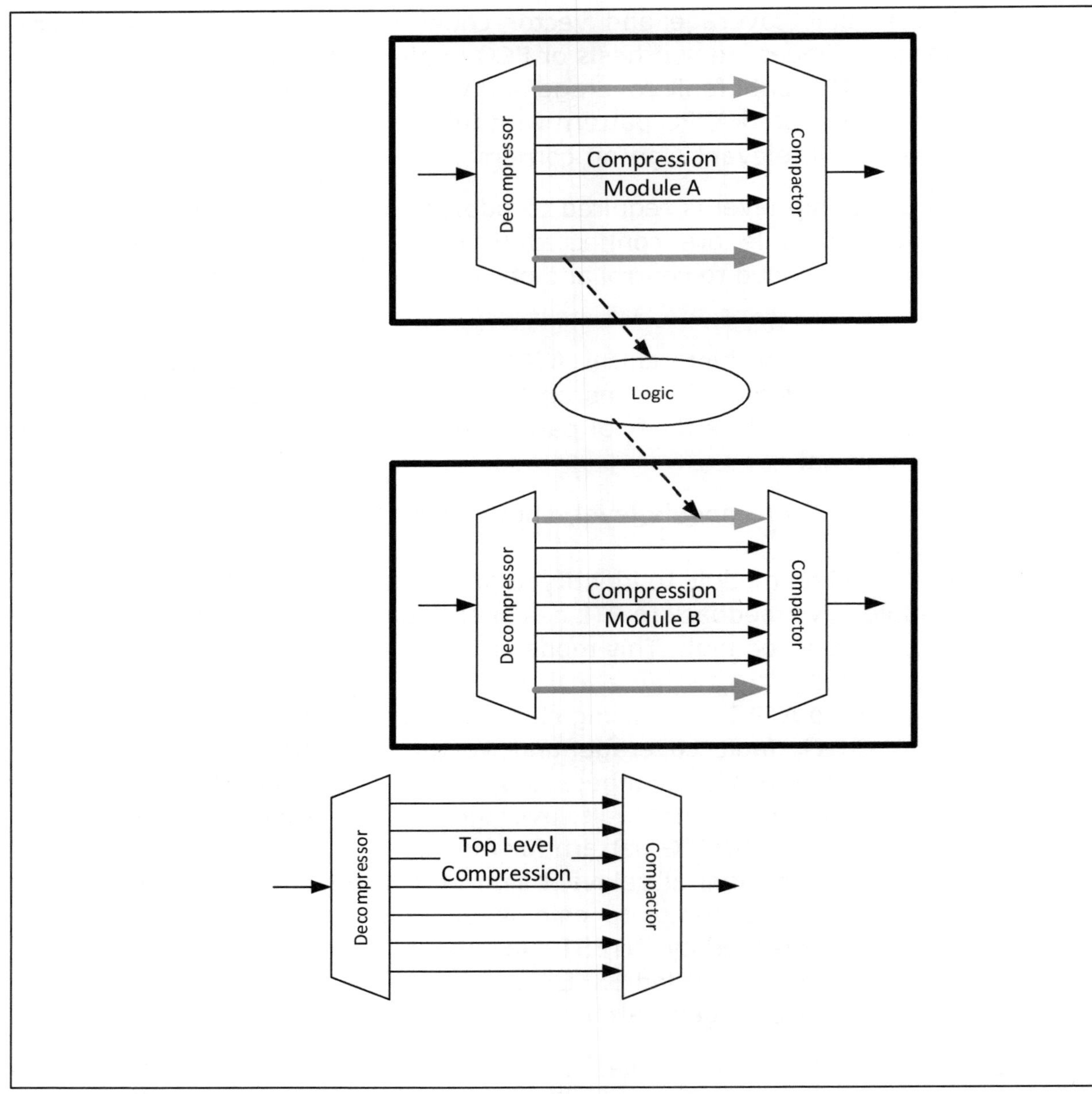

Figure 53: Modular compression enhanced for hierarchical ATPG

The mode switching circuitry allows to overcome the problem of uncontrolled and unobserved logic outside modules. The scan system can be configured in such a way that all flip-flops driving logic outside the module can be connected as one or more internal scan chain(s). Flip-flops receiving data from logic outside the module are configured in the same way within dedicated internal scan chain(s).

Using all wrapper scan chains for SoC-level ATPG allows full control and observe via scan flip-flops without considering all flip-flops inside the hierarchical modules. Thus, ATPG can be done for module A and B as soon as these are available (INTEST runs). Redo of ATPG is required only in case of changes for the respective module. To allow full access via scan for all top-level logic red and green scan chains from module level compression systems are included into top-level ATPG (EXTEST runs). This requires besides configuration of the wrapper scan chains some additional logic inside compression and a special consideration for scan clock circuity.

A modular compression system enhanced for hierarchical ATPG can be used beneficially for following purposes:

1. Shorten ATPG flow by doing block-wise pattern generation. A finished block can be run through ATPG and gate level simulation. Generated patterns can be merged within a post-processing step with pattern for other blocks and top-level ('Retargeting'). This usage enables significant productivity improvements for the ATPG flow. All partitions (Module A, Module B and Top-Level in Figure 53) can run ATPG in parallel.
2. Compression channels can be assigned to GPIO supporting multiple scan modes easily. As an example, related to Figure 52 and Figure 53 'Scanmode 1' execute scan test for modules A and B together. 'Scanmode 2' may run top-level ATPG pattern only. Such an architecture needs to be considered during RTL design and can overcome limitations in number of available scan port signals.

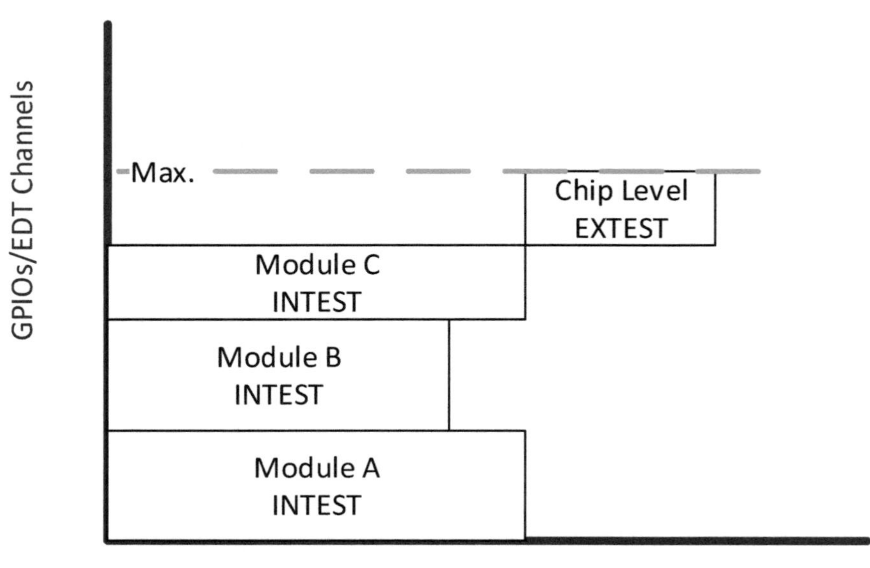

Figure 54: Test execution sequence for hierarchical ATPG

Figure 54 shows a possible sequence of ATPG vectors during scan test. As in conventional, non-hierarchical scan each module including chip level gets its own set of GPIO. Module level ATPG patterns from INTEST runs are executed concurrently. Since peripheral scan chains are used twice for INTEST and EXTEST chip level ATPG patterns have to be executed separated from all INTEST runs as shown.

Since EXTEST patterns need to be separated from INTEST anyhow GPIO assignment can be optimized as shown in Figure 55. GPIO assigned to chip level are freed-up and distributed for module level test. This allows a modest reduction of test time for INTEST runs. EXTEST is using all available GPIO and test time can be reduced drastically for chip level runs.

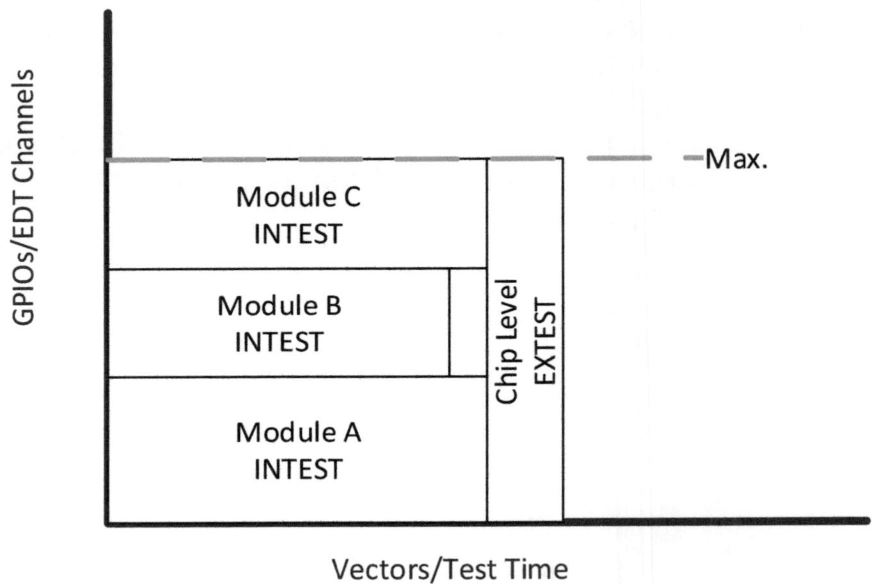

Figure 55: GPIO assignment optimized for hierarchical scan

What is the impact on test cycles for hierarchical scan approach? Let's start with the basic formula for test cycles N:

$$N = V * F/GPIO$$

For hierarchical scan flip-flops inside wrapper chains are loaded twice during ATPG, for INTEST and EXTEST modes. In conventional scan this is not required. Thus, we have to consider F representing all flip-flops and flip-flops in wrapper chains F_{wrap} additionally:

$$N = V * (F + F_{\mathrm{wrap}})/GPIO$$

On the benefit's side the ATPG tool is able to provide a reduced set of vectors since circuit size and by this ATPG problem size is smaller. ATPG tool vendor claims a reduction of vectors V by 2x or 3x. The resulting formula considering such reduction factor v_{red} is as follows:

$$N = V/v_{\mathrm{red}} * (F + F_{\mathrm{wrap}})/GPIO$$

To evaluate a conservative case let us assume 50% of flip-flops are wrapper flip-flops and 1.5x vectors less are generated. This results in $v_{\mathrm{red}} = 1.5$ and $(F + F_{\mathrm{wrap}}) = 1.5$. For this example, test cycle-wise use of hierarchical scan is neutral. If number of wrapper flip-flops is less and/or vector reduction is higher, overall number N of test cycles is reduced. The reported vector savings are higher [12].

Optimizing modular scan and hierarchical ATPG

Figure 56 indicates a single fault (marked as asterisk) targeted by ATPG. On the left side scan chains above and below are empty and show the status before ATPG. The right side shows state of flip-flops inside scan chains after ATPG is finished for the indicated fault. All 'X' are interpreted as don't care values, i.e. those flip-flops are not needed to detect the marked faults. In this example let us assume two flip-flops containing a '1' state are needed to control (initialize) the fault and the fault effect is observed (captured) via two flip-flops marked with 'D'. 'D' indicates a logical difference between combinational logic containing the marked fault and the fault-free state. The portion of flip-flops not marked with 'X' is called the fill rate for that fault. A medium fill rate is in the range of 1-2%. Scan compression makes use of this low fill rate by targeting more faults with a single scan chain loading. By this scan vectors are compressed. As a consequence, compression is also limited to 1/1%, i.e. 100x. To increase compression rates significantly above this results in ineffectiveness of the ATPG process. Therefore, ATPG vector count increases to levels compensating scan compression gains.

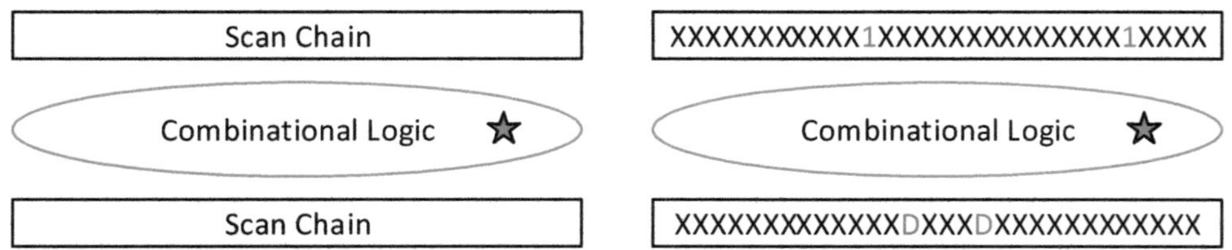

Figure 56: ATPG use a small subset only for a single fault

Defining compression architecture for given partition should start with a compression factor of ~100x as first parameter to be selected. Based on the number of flip-flops within the partition and the aligned chain length the number of internal chains is given. Required external chains need to be selected such that the ratio of external to internal channels results in ~100.

Generally, choosing the optimum partition size **conflicting** parameters have to be considered:

- For minimizing ATPG run time: make **partitions as small as** possible.
- Flip-flops within wrapper chains have to be loaded twice: for addressing faults inside the partition (INTEST mode) and for faults in top level (EXTEST mode), this double loading requires vector memory on ATE as well as test time. Consequently, the overall number of **wrapper cells should be as low** as possible.
- For minimizing additional logic required to control partitions **make no partitions** at all:
    - Clock, reset add-ons
    - Wrapper chains: insert dedicated wrapper cells to cut boundaries

Optimizations used commonly are asymmetric scan compression and input channel sharing. Both techniques are described below.

Using asymmetric compression

Basic idea is to use different numbers of EDT channels for input and for output. This results in individual compression rates for input and output sides. Based on experiences 2-3x more channels can be used for inputs than for outputs without increasing the number of required ATPG vectors. Overall result is an improved bandwidth for input side and by this an overall reduction of scan pattern size stored on ATE.

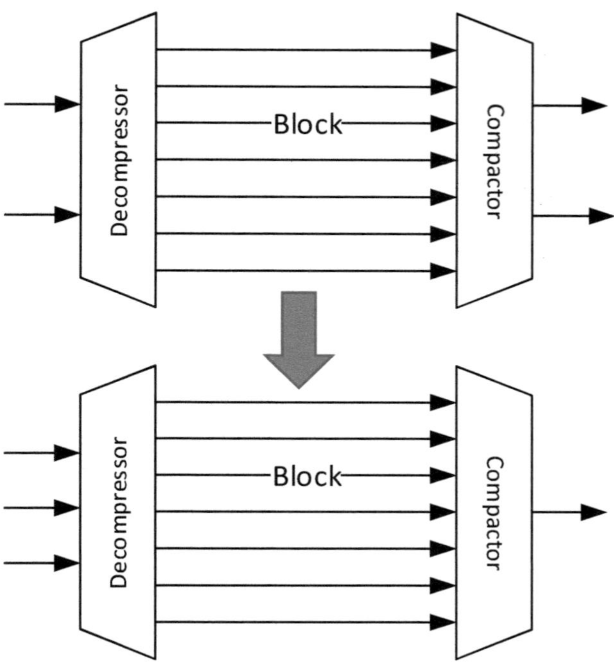

Figure 57: From symmetric to asymmetric compression - using more input that output channels

The example shown in Figure 57 converts one output channel into an input channel. By this, more inputs can be used. More and shorter scan chains are possible with same compression rate for inputs.

Channel Sharing for Inputs

Basic idea for channel sharing originates from stimulating identical modules using the same input channels as shown in Figure 58. Combining two identical modules Block A results in 2x higher input compression rate for these two blocks.

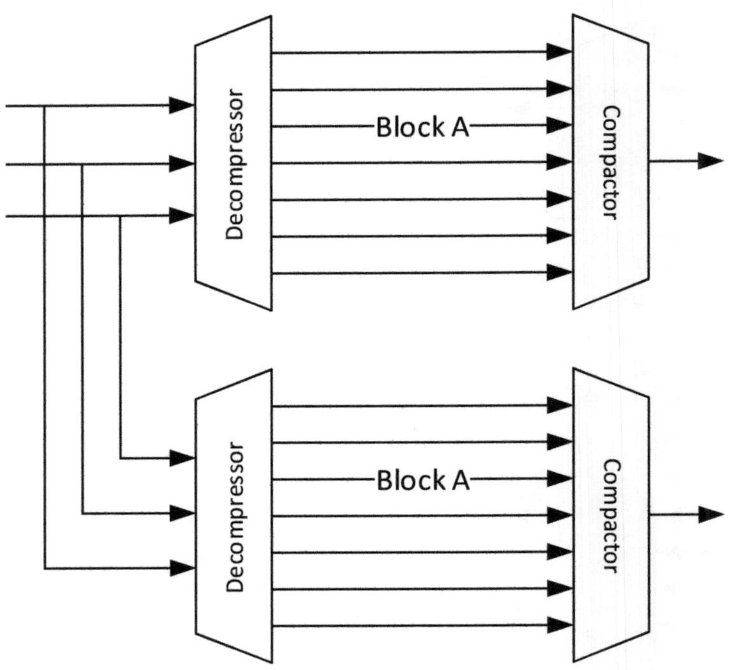

Figure 58: Channel sharing for identical modules

Extending the basic idea to non-identical modules is shown in Figure 59. At least one channel needs to be unique for each module. These channels are named as control channels (control A and control B) in Figure 59. Channel sharing is possible for other channels, referred to as data channels. Input channel requirement for the example from Figure 59 is reduced from 6 to 4 channels combining Block A and Block B. Input compression rate is improved by 6/4 = 1.5.

If partitions to enable hierarchical ATPG (see Chapter Hierarchical ATPG) are used channel sharing is possible within a single partition only. Otherwise ATPG is not able to provide identical vectors for data channels.

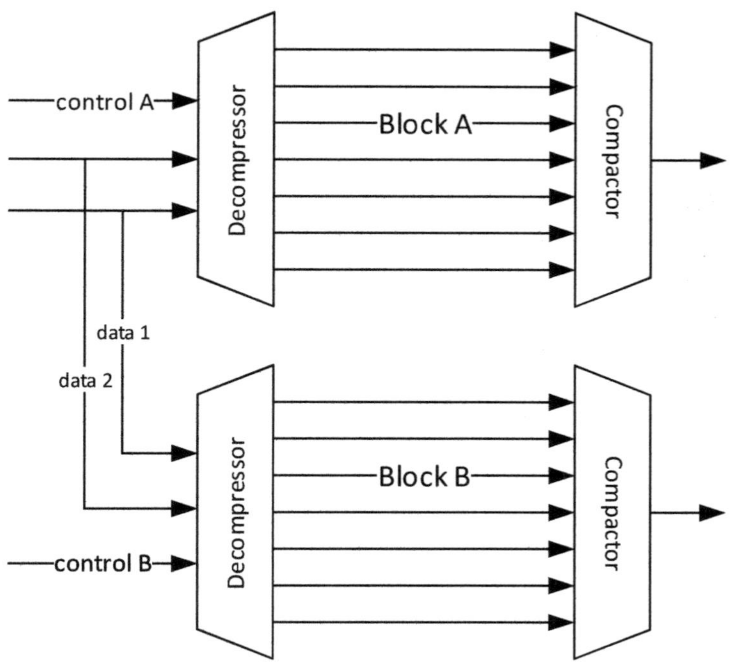

Figure 59: Channel sharing for two different modules with control and data channels

The influence of number of scan modes

The concept of scan partitions allows the grouping of partitions into different scan modes. For each scan mode input and output channels of a selected set of partitions are connected to GPIO (General-Purpose-Inputs-Outputs)[5]. For the design in

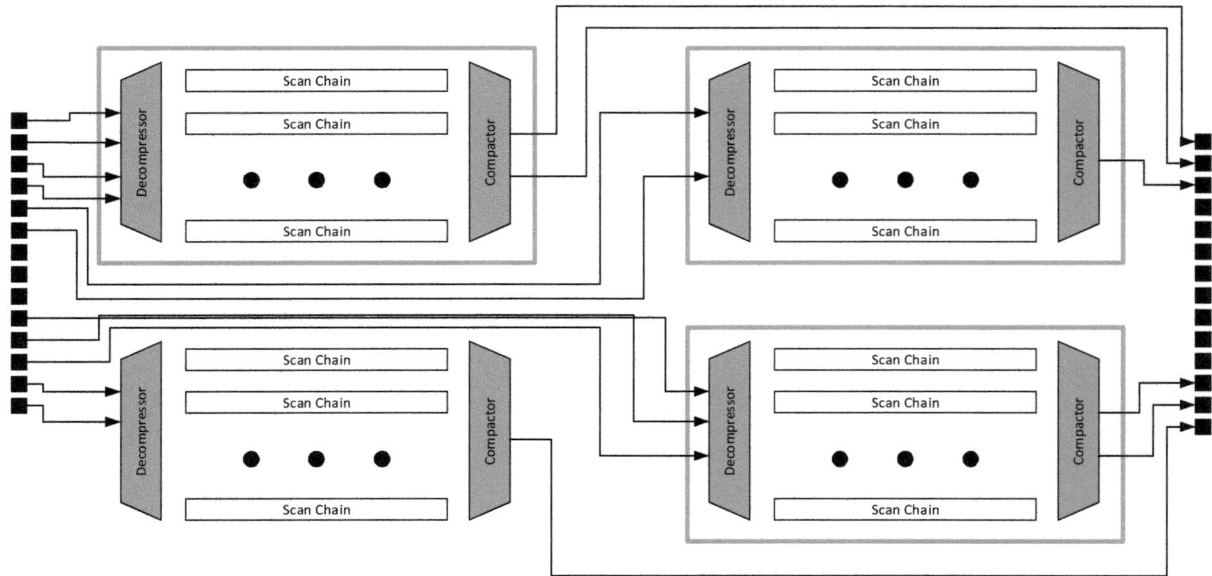

Figure 60: System-on-Chip with 4 scan partitions

---

[5] GPIO are used for various test and functional modes and can be configured typically individually as input or output

enough GPIOs (black rectangles at left and right side) are available to connect all input and output channels. Here only a single scan mode is required.

Now let's assume we have less GPIO available forcing us to use two scan modes:

- Scan Mode 1: top two partitions, requiring 6 input and 3 output channels
- Scan Mode 2: two partitions at bottom, requiring 5 input and 3 output channels

The vector memory at ATE for each GPIO has to store scan data for both scan modes. To get the time for scan test execution the sum of both scan modes has to be used.

Scan bandwidth comparison

Here we define 'scan bandwidth' to allow a comparison of different scan architectures. With this approach we can separate the influences of scan architecture, the ATPG effectiveness and the circuit complexity. The assumption is that the number of generated vectors N (i.e., ATPG effectiveness) is independent of scan architecture and circuit complexity. This has been proven as valid within ranges of interest for current products.

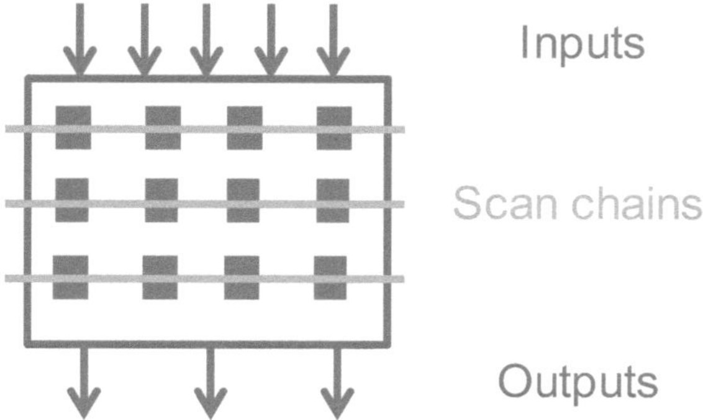

Figure 61: Basic scan architecture

As a reminder the basic scan architecture is shown in Figure 61. The basic formulas for test cycles, test time and ATE test data volume are as follows:

Test cycles = N * l
Test time = 1/f * N* l
ATE test data = N * l (to be stored per ATE channel or per scan chain)

Hereby N is the number of vectors provided by the ATPG tool representing a certain fault coverage. The length of (longest!) scan chain is l and f the frequency applied during scan test. For this example, we can introduce a basic definition for scan bandwidth:

b = GPIO/2 * f

With GPIO representing the number of available I/O for scan chains. The scan bandwidth is the number of scan chains multiplied by the applied scan frequency. The test time formula is now:

Test time = (N * F)/b with the number of scan flip-flops F.

The term (N*F) represents the data volume summed up for all channels which need to be stored on ATE and applied via the scan interface. The capability of the scan port is represented by bandwidth which is a product of width (here the number of scan chains) and frequency of scan shift operation.

For scan architectures using on-chip decompression and compaction (Figure 62) the data volume is reduced by compression factor r:

$$\text{Test time} = \frac{\text{Data volume}}{\text{bandwidth}} = \frac{N * F}{r} * \frac{1}{b}$$

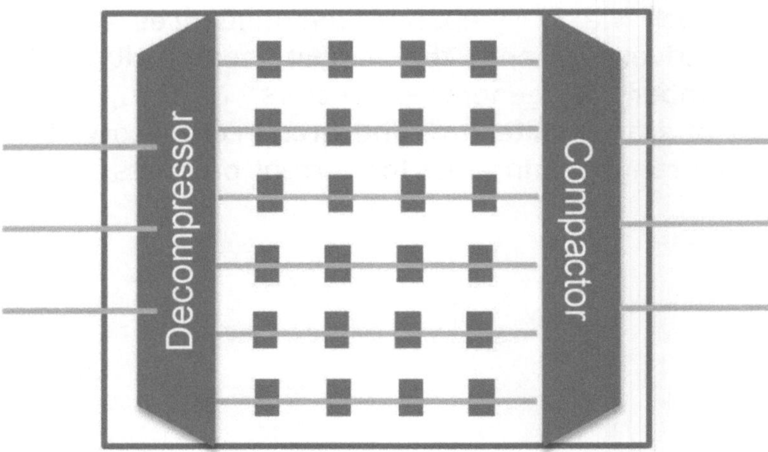

Figure 62: Scan compression

The scan bandwidth is the applied frequency multiplied by the number of EDT channels. A single EDT channel is driving r-times the number of internal scan chains.

The chosen definitions allow a clear separation of influencing factors for the test time:

- N (number of scan vectors or scan loads) gives efficiency of ATPG algorithm and software implementation
- F (number of flip-flops) gives circuit's internal complexity
- b (bandwidth[6]) is measure for ATE-to-scan interface
- r (compression) gives the influence of internal logic for data reduction

A hierarchical scan architecture (principle shown in Figure 63) can have different compression factors for partitions. Data volume per partition is given by:

$$Vp = \frac{Np * Fp}{rp}$$

The total data volume is calculated as sum over all Vp. If all partitions are tested within a single test session, i.e. a single configuration of scan port to EDT channels, test time is calculated as follows:

$$Test\ time = \frac{V_{\text{EXTEST}} + \sum_P Vp}{\text{EDT channels} * f}$$

V_{EXTEST} gives the data volume required for the EXTEST mode.

---

[6] this fits to the common understanding of digital bandwidth as measured in bits per time unit

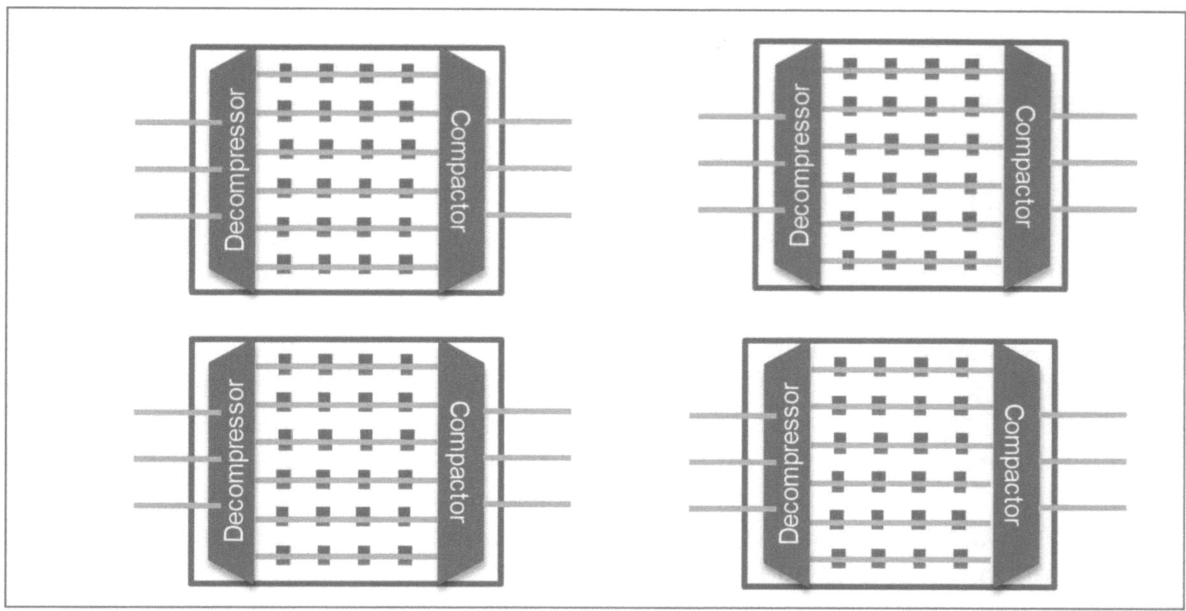

Figure 63: Hierarchical scan architecture

For asymmetric compression architectures the input and output compression ratios are different. The experience shows this is almost neutral to vector count N if $r_{in}/r_{out} < 4$. For data volume reduction r_{in} needs to be used.

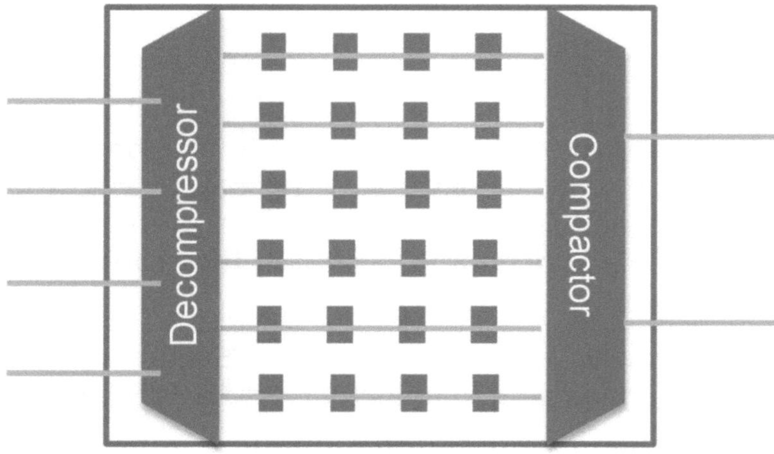

Figure 64: Asymmetric compression

Memory Test and BIST

Regarding the layout of SoC embedded memories are typically visible as most regular structures. A certain number of cells is organized in horizontal and vertical directions. Nowhere in the layout are distances between transistors closer. This narrow neighborhood allows compact realization of embedded memories. The close physical distance between transistors needs more thorough testing and more advanced faults models than used for standard cell structures.

Classification of memories

The dominant types for SoC are SRAM (Static Random Access Memories) and ROM (Read-Only Memories). There are also specialized manufacturing technologies which allow DRAM (Dynamic Random Access Memory), Flash or RRAM (Resistive Random Access Memory) as embedded structures. As commodity products DRAM and Flash are highest volume microelectronic products. Generally, memories can be volatile or non-volatile. Volatile memories are losing their information if power supply is switched-off. DRAM and SRAM types belong to the class of volatile memories. Non-volatile memories keep their information even if power is no longer supplied. A classification scheme is shown in Figure 65.

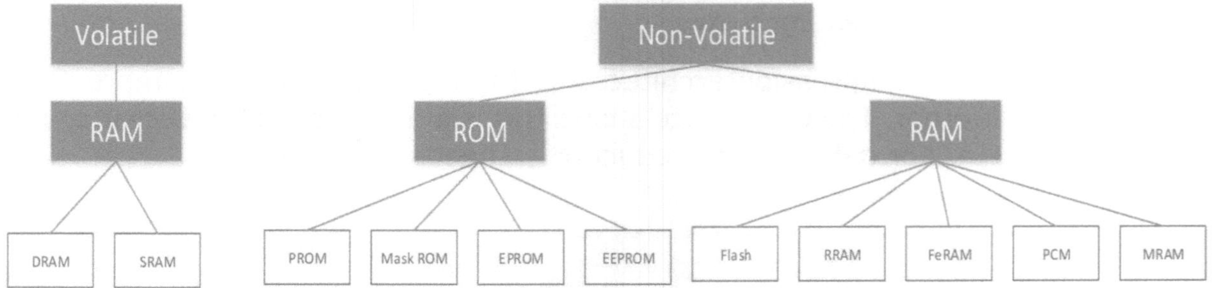

Figure 65: Classification of memory types

A second classification level are random-access and read-only memory type. Random-access memories allow to read, modify and write data during the normal operation. Read-only memories provide data which are written once. This takes place during Silicon manufacturing for ROM or after time consuming erase steps like for (E)EPROM types.

Since memories are the most densely packed structures for SoC they need special consideration for manufacturing test and are important for feedback to the manufacturing technology. The regular structures are most sensitive to defects.

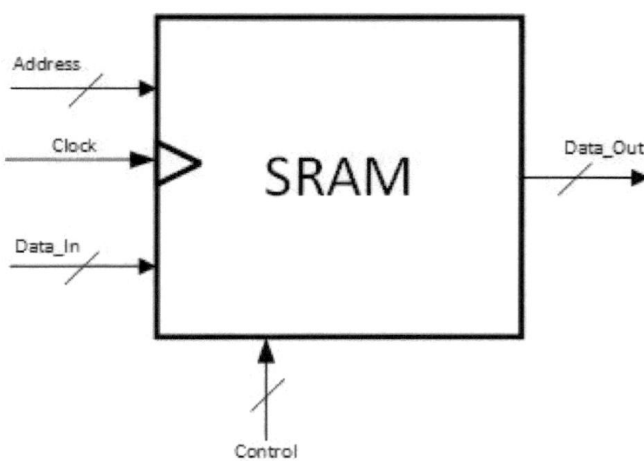

Figure 66: SRAM interface signals

Figure 66 shows the input and output lines for a single-port SRAM with uni-directional interface. The SRAM is organized in data words of a certain width. The data width gives the number of required lines for Data_in and Data_out busses. Data words are accessible via addresses which are provided via Address bus according Figure 66. The required address bus width is log2(number of data words), e.g., 10 address lines for 1024 words. Clock and some control lines handle operation of the SRAM.

There are variations used: Data_In and Data_Out are combined into a bi-directional data bus and access via two or more data ports is possible.

The internal building blocks for an SRAM are shown in Figure 67. Area-wise the cell array is the dominating part. Here the information is stored. The addresses are selected via Address, Column and Row Decoder blocks. Data are read via Sense Amplifier block and available at the Data Register. Write operation uses the Data Register and the Write Driver block.

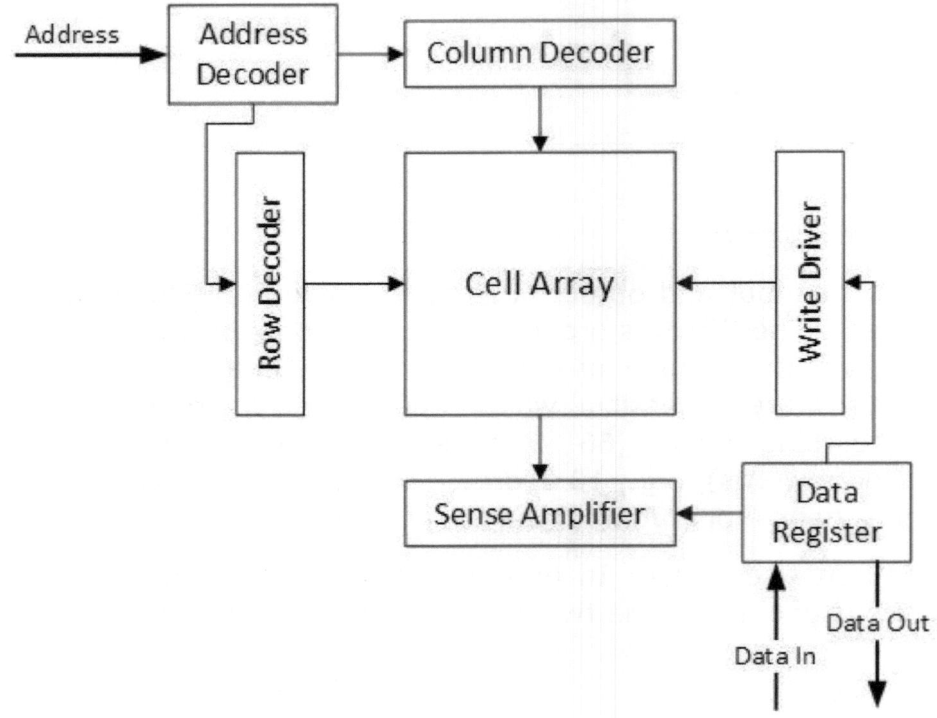

Figure 67: Structure of an SRAM block

For DRAM additional circuitry is required which refreshes the content regularly. Re-programmable memory types require logic to erase the content. The cell circuitry is specific for a certain memory type. A DRAM cell consists of a single transistor plus a lateral capacitor to store the information. An SRAM cell requires six transistors, and the information is stored using two cross-coupled inverters as shown in Figure 68.

Memory fault models

Compared to standard cell-based logic additional faults have to be considered for memories. The most used fault models for memory testing are:

- Cell stuck-at
- Cell transition
- Cell data retention
- State coupling within the array
- Pattern sensitivity within the array
- Stuck-at within periphery blocks

Based on these faults algorithmic tests are constructed which check the array for defects. To explain the listed fault models state diagrams are used.

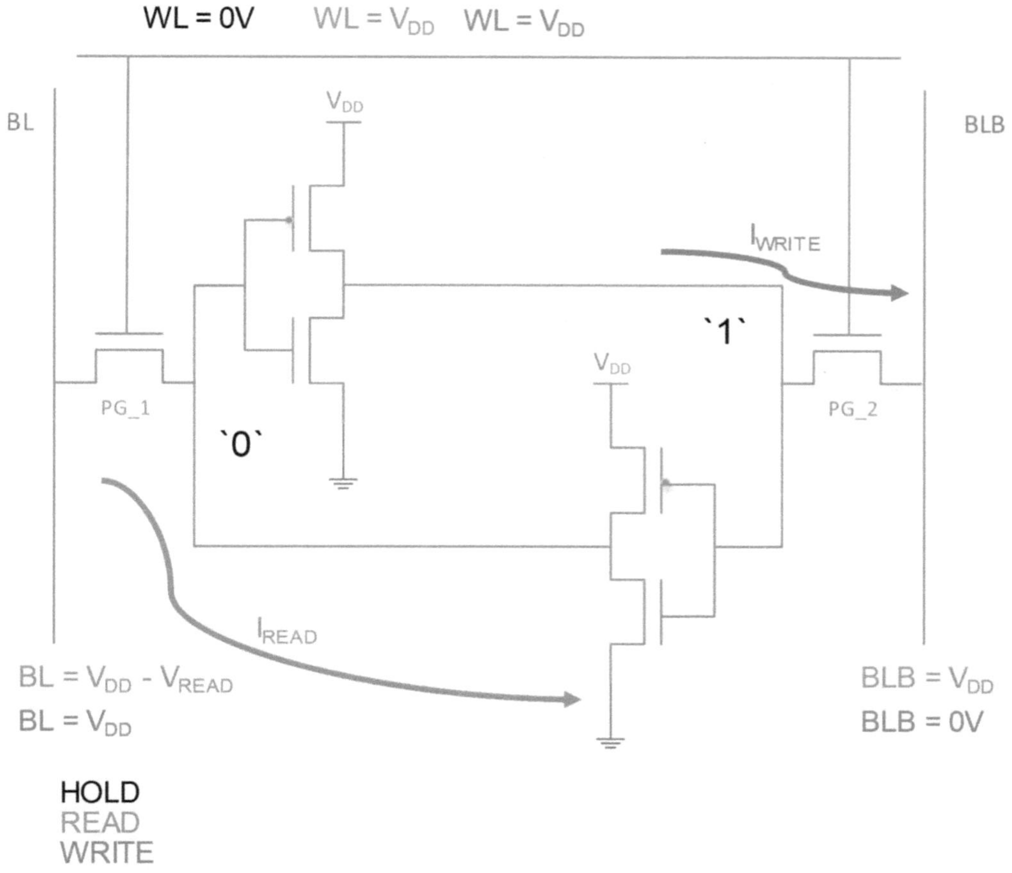

Figure 68: SRAM cell using cross-coupled inverters

Figure 69 shows fault-free and faulty state diagram. A memory cell can have two states depending on if '0' (S0) or '1' (S1) are stored. During read access states are not changing their value. A write '0' command changes S1 and leaves S0 unchanged. Stuck-at behavior is shown in the middle of Figure 69. Independent of the command applied the cell's state does not change. A state diagram for possible transition faults is shown at the bottom. Write commands can result in wrong cell states and read commands are able to change the state in presence of a transition fault.

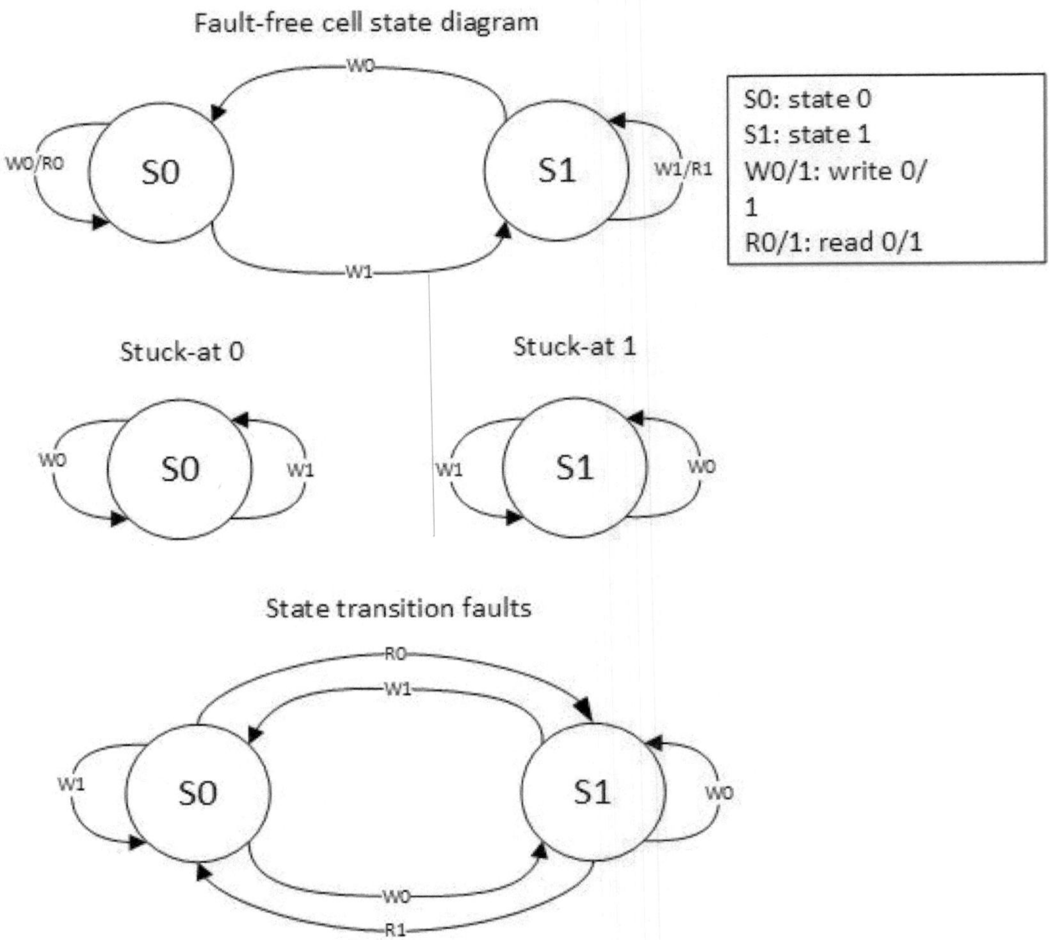

Figure 69: Cell state diagrams for fault-free, stuck-at and transition faults

Memory arrays have closest distances for transistors. This results in a certain probability for coupling faults between neighbored cells. If a cell changes its value, the neighboring cell changes also its value (Inversion coupling fault in Figure 70). An idempotent coupling fault results in the same value for the targeted cell and the neighbored cell. A pattern sensitive fault influences a victim cell depending on a certain pattern (value set) within all neighboring cells.

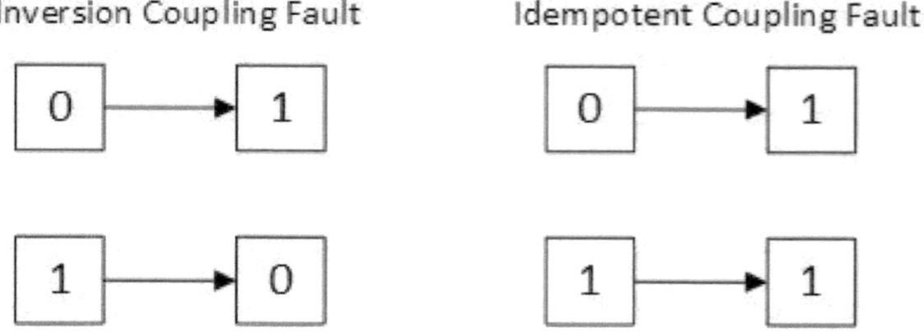

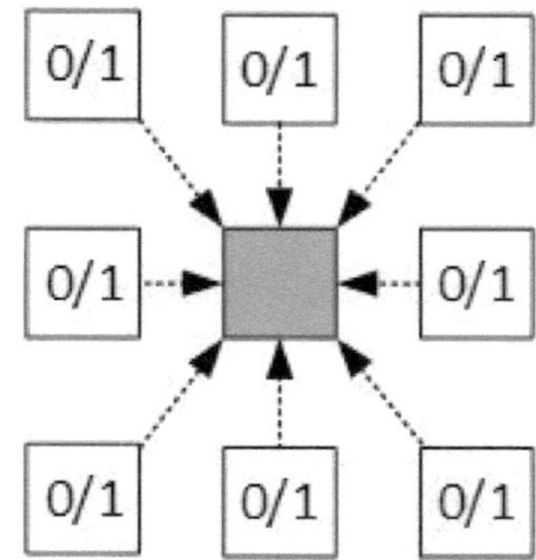

Figure 70: Coupling faults

Memory test algorithms

For testing memories background patterns are an essential part. Checkerboard and row bar background patterns are often used and shown in Figure 71. A background pattern is not sufficient to test the complete array. Assume row bar is used and be reminded that I/O experiments are used to test the cell array. Writing row bar into the array and reading back all '0' and all '1' only guarantees that two addresses are working correctly. There is no confirmation that all addresses are working.

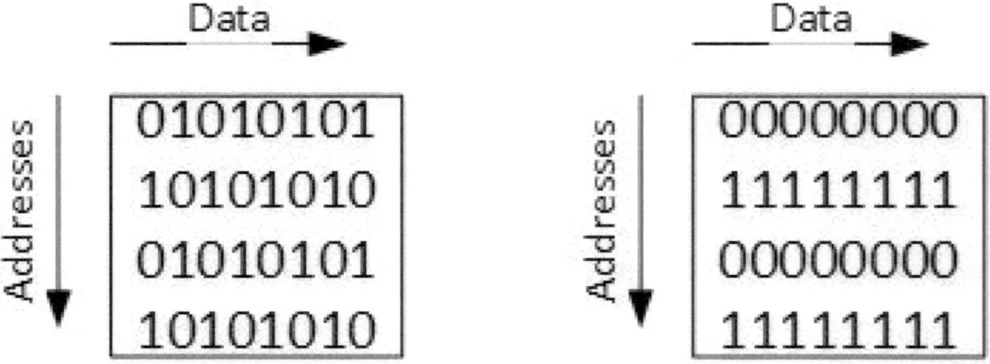

Figure 71: Checkerboard and row bar background pattern

To prove that all addresses are accessible March tests are used. An example for a March algorithm is shown in Figure 72. The algorithm starts with the lowest address writing a '0' into all cells. This can be regarded as an 'all-zero' background pattern. Accordingly, the second line of the algorithm description the algorithm checks the first address if the value '0' is already present, writes a '1' into the first address and checks if the '1' is readable. Afterwards the algorithm proceeds with the next address, executes read '0', write '1' and read '1' sequence until the highest address is reached. This can be seen as a '1' marching through the array. For the next line the algorithm starts again with the lowest address, executes the read-write-read operation for each address. Every line is executed for all addresses. The second part of the algorithm runs the array from highest to lowest address. The sum of all read and write commands gives the complexity of the algorithm. For this example, all addresses are accessed 14 times. This gives the time complexity for the

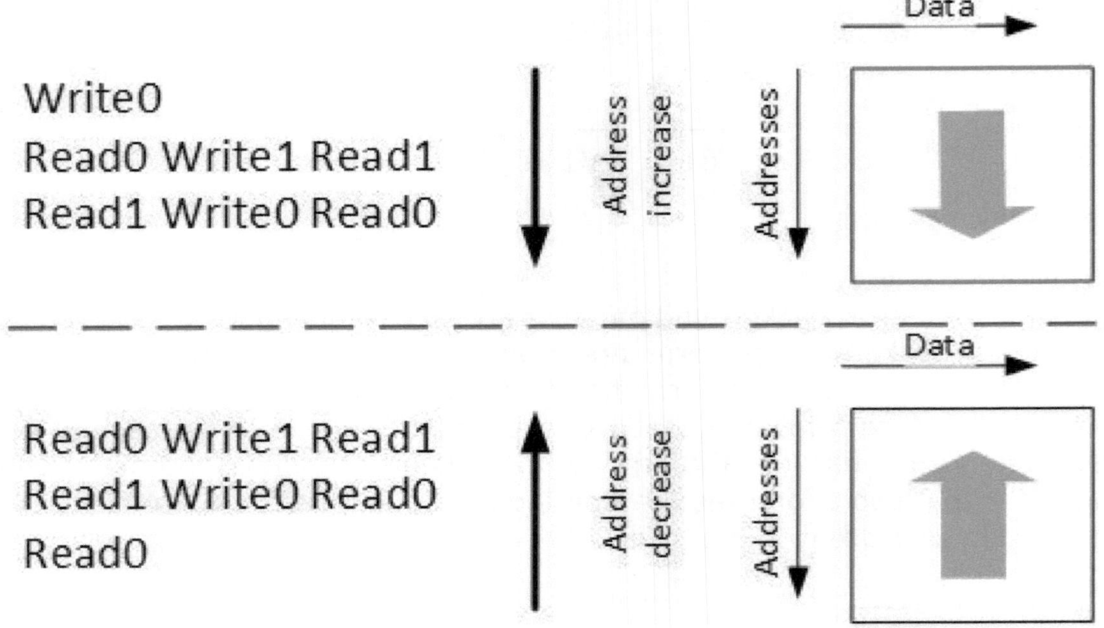

Figure 72: March-14N test algorithm

algorithm which is 14N. The complexity measure N gives the number of addresses. For the selected example this is identical to the number of cells in the array (bit-oriented memory).

Using a very common notation as used in [22] describes the March-14N as follows:

$$\{\Uparrow(w0); \Uparrow(r0, w1, r1); \Uparrow(r1, w0, r0); \Downarrow(r0, w1, r1); \Downarrow(r1, w0, r0); \Downarrow(r0)\}$$

To address especially pattern sensitive coupling faults so called Galpat algorithms are used. An example is shown in Figure 73. The command 'Read(all other cells) 0/1' is interpreted as reading all cells of the array, which can have a '1' or '0' value. The second line in the algorithm description starts with reading '0' for the first address followed by writing a '1' into the first cell. After the write command the whole array is checked if all expected values are still unchanged ("Gallopping through the array"). The commands for the whole line are repeated for the next address. Such an algorithm no longer has a linear complexity. For the example of Figure 73 the complexity is N+2*N*(2+N-1) = 3N + N^2. Using concrete numbers for a 4 Mbit memory tested at 100ns (10 MHz) results in 55 days for the Galpat and 5.6s for the March 14N algorithm. For embedded SRAM test March algorithms are used as industry standard.

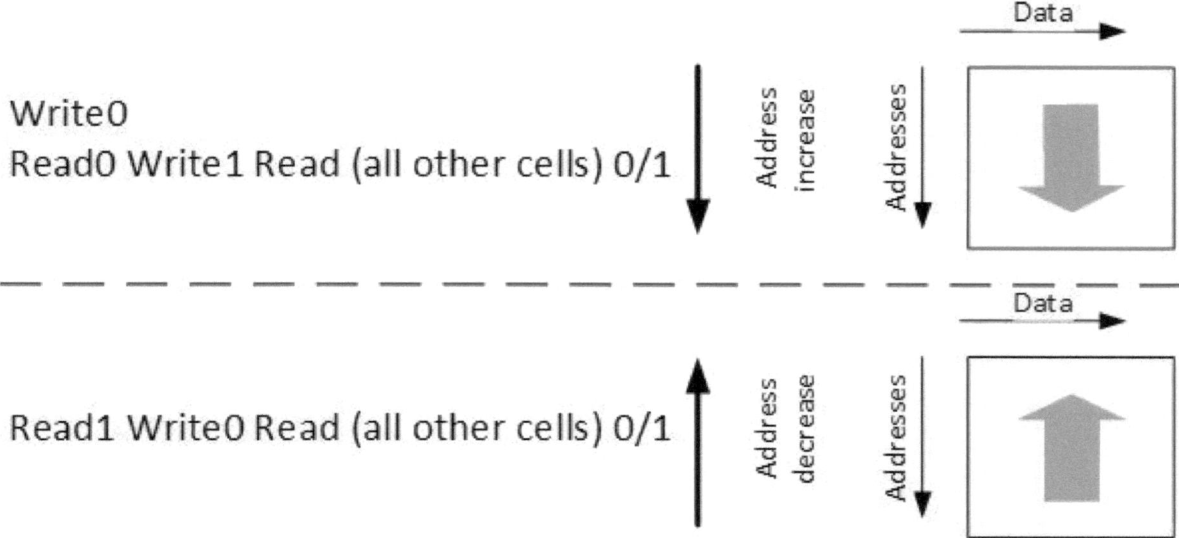

Figure 73: Galpat algorithm

Memory BIST

A typical SoC contains up to hundreds or even thousands of embedded memory instances. To route all interface signals via the I/O structure to the ATE would require a lot of resources. A single 32-bit wide SRAM with a capacity of 1000 addresses would require direct access by ATE of 2*32 data lines plus 10 address lines and some control signals.

To overcome this BIST is a perfect candidate. BIST logic close to the memories allows direct application of the algorithm and pass/fail decision at the memory outputs. Another benefit spending area for BIST circuitry is the ability to execute memory test concurrently.

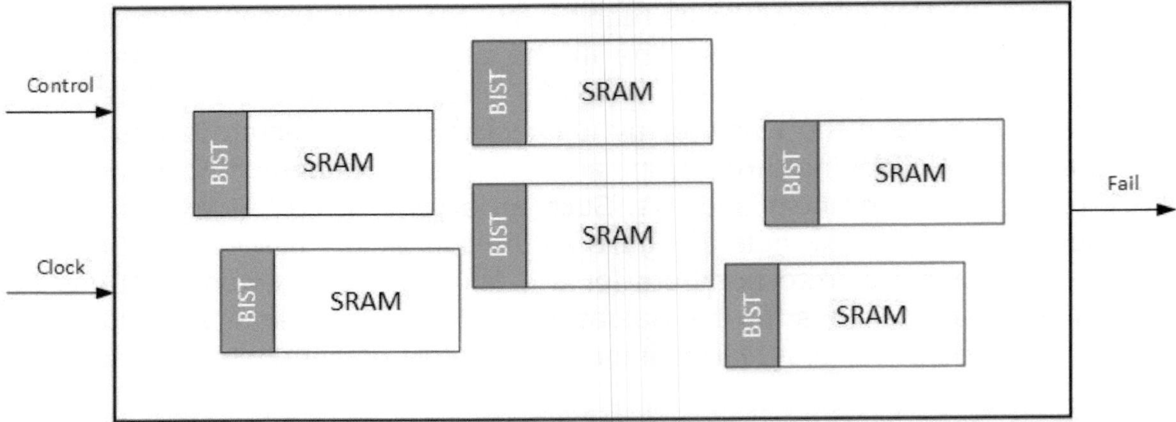

Figure 74: Concurrent test of embedded SRAM instances using BIST

Figure 74 shows the principle of concurrent embedded SRAM test. The ATE provides clock and control signals. The fail signal for the BIST runs is evaluated at the ATE. The on-chip power network limits how many SRAM instances can be tested in parallel. Typically, the memories are split into multiple groups where a single group is tested concurrently. The largest array for each group dominates the test time.

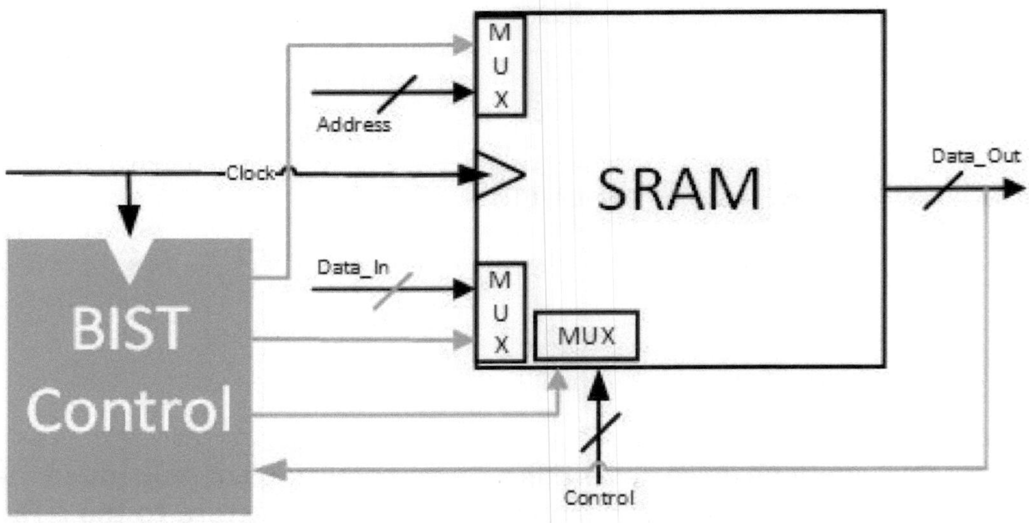

Figure 75: Embedded SRAM with BIST circuitry

BIST circuitry for SRAM can be realized using compact logic, see Figure 75. Since algorithms are used a generator can be constructed using finite-state machines. Expected data at memory outputs are known for every cycle and can be compared directly at the SRAM's output. Multiplexers are needed to control the interface during test and a local BIST controller is required to execute the test operation.

Redundancy and repair for memories

If embedded memories cover a significant area of SoC yield loss related to memories becomes dominating. Assume logic and memory both have same area portion of 40%, yield loss by memory tests is higher originating from higher defect sensitivity. To recover some or all of the memory-related yield loss repair schemes are available. Embedded memories are implemented with redundant rows or columns which are activated in case of a defect inside the memory array. An overview about different repair techniques is given in [25].

The optimization criteria for redundancy and repair is good-dies-per-wafer. Implementing redundancy increases the die area and results typically into less dies realized on a wafer. Redundancy pays-off if the number of resulting pass devices for a wafer after repair is higher than without redundancy.

Diagnosis of memory fails

Embedded memory structures are of special interest for manufacturing technology based on narrow transistor distances. A lot of systematic issues which arise during wafer processing can be diagnosed with the help of memory bitfail maps. At least for a subset of all memory instances a debug option must be available to provide more information than just pass/fail. During debug runs all fails for an array are collected and bitfail maps are constructed out of the debug information. Examples for different bitfail pattern are shown in Figure 76.

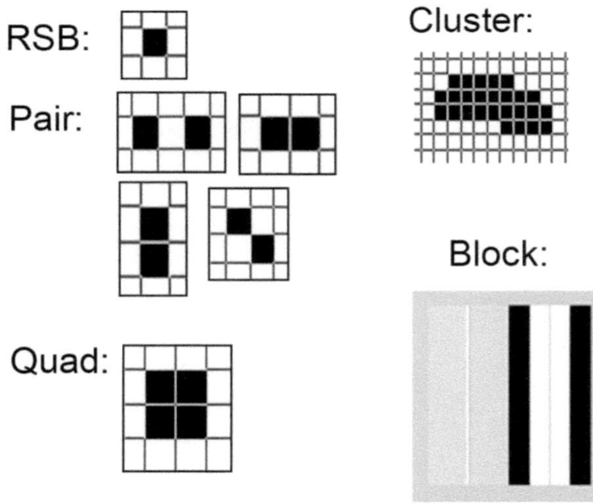

Figure 76: Examples for bitfail maps

BIST for other Structures

The implementation of logic for LBIST and MBIST is supported by EDA tools and follows typically a structured and scalable architecture. For some special functions there are also BIST schemes known. These are dedicated to the respective function. Prominent examples are BIST for PLL and loopback for HSIO interfaces. Here the BIST function is added by the module designer and no EDA tool is available.

The Chapters in Test and DfT for special structures contain also descriptions of the BIST functionality.

Boundary-Scan and JTAG test access port

The boundary-scan standard IEEE 1149 is one of the eldest and most prominent DfT standards. The first version was released in 1990, the latest update is from 2013 [26]. The main motivation for this standard was to support the test of printed-circuit boards (PCB). This on-chip DfT measure should help to ease the test of densely packed, multi-layer PCB after mounting integrated circuits on the board. By this PCB interconnect as well as soldering defects could be tested more efficiently.

Besides supporting PCB, test elements of this standard are used for test control and manufacturing of I/O structures. The main idea is to provide access between core logic and pads of integrated circuits.

Figure 77 shows boundary-scan cells (squares) between core logic and device's inputs and outputs. These cells are serially connected to form the boundary-scan register. The serial input to the boundary-scan register is a dedicated TDI pin and a serial output is provided via TDO pin. Other elements of the boundary-scan logic are a TAP-FSM, instruction register and instruction decoder. TAP-FSM is a finite-state machine controlling the test access port (TAP). TMS, TCK and TRST* are 3 additional pins required to control the boundary-scan system. The TAP is able to connect other registers than the boundary-scan register between TDI and TDO. Defined by the standard are bypass and device identification register. Additionally, an arbitrary number of design-specific registers can be controlled by the TAP. For a typical SoC TAP-FSM, instruction register and decoder, bypass and device ID register, TDO and register multiplexers are combined into a TAP module. Together with the 5-pin interface this is also called JTAG[7] test control interface.

---

[7] The creators of the first version of IEEE 1149.1 standard named themselves Joint-Test Action Group

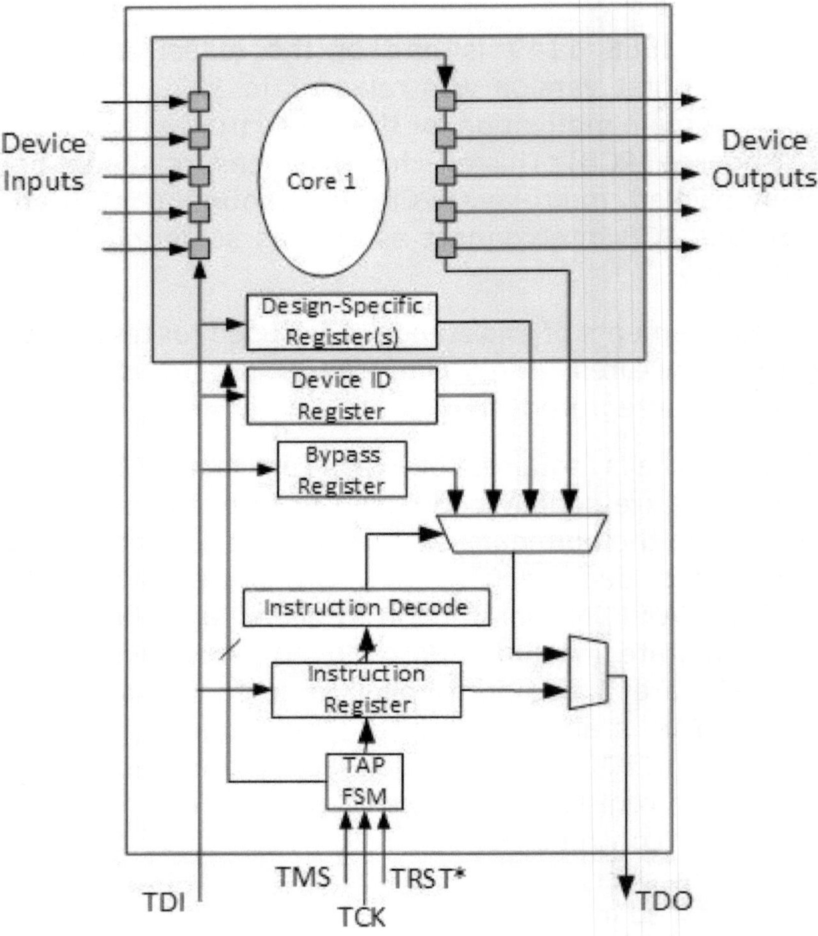

Figure 77: Boundary-scan logic

The following table lists the interface pins, which are not allowed to be shared with other functions according to the standard.

Name	Function	Comment
TDI	Test Data In	Serially shifts in data into the device on rising edge of TCK according to selected state of FSM
TMS	Test Mode Select	Determine next state of FSM on rising edge of TCK
TDO	Test Data Out	Serially shifts data out of the device on falling edge of TCK according to selected state of FSM
TCK	Test Clock	Clock for complete boundary-scan system: FSM, instruction register, and all test data registers
TRST*	Test Reset	Optional according standard: provides asynchronous, low-active reset for complete boundary-scan system

The registers defined by the standard are summarized with the next table. According to the standard all registers, except the instruction register, are test registers.

Name	Length	Comment
Instruction	Any	Single instruction for every register, bypass register can be accessed by multiple instructions
Boundary-Scan	Any	Between core logic and pad circuitry
Bypass	1bit	
Device ID	32bit	Defined format, central administration, register optional by the standard
Design-specific	Any	No limits regarding number

Operation of the TAP-FSM is defined by the state transition diagram given by Figure 78. All state transitions are synchronized with TCK and controlled by TMS value. Operation of the FSM starts after reset walking through states for the instruction register. The related states are in the right branch of the diagram containing an 'IR' in the name. During state 'Shift-IR' the instruction code is shifted in serially into the instruction register. Entering state 'Update-IR' the loaded code is transferred into the parallel output stage of the instruction register. With this a single test register is selected. Using the states of the middle branch in Figure 78 allows operations with the selected test register. The most important states are 'Capture-DR', 'Shift-DR' and 'Update-

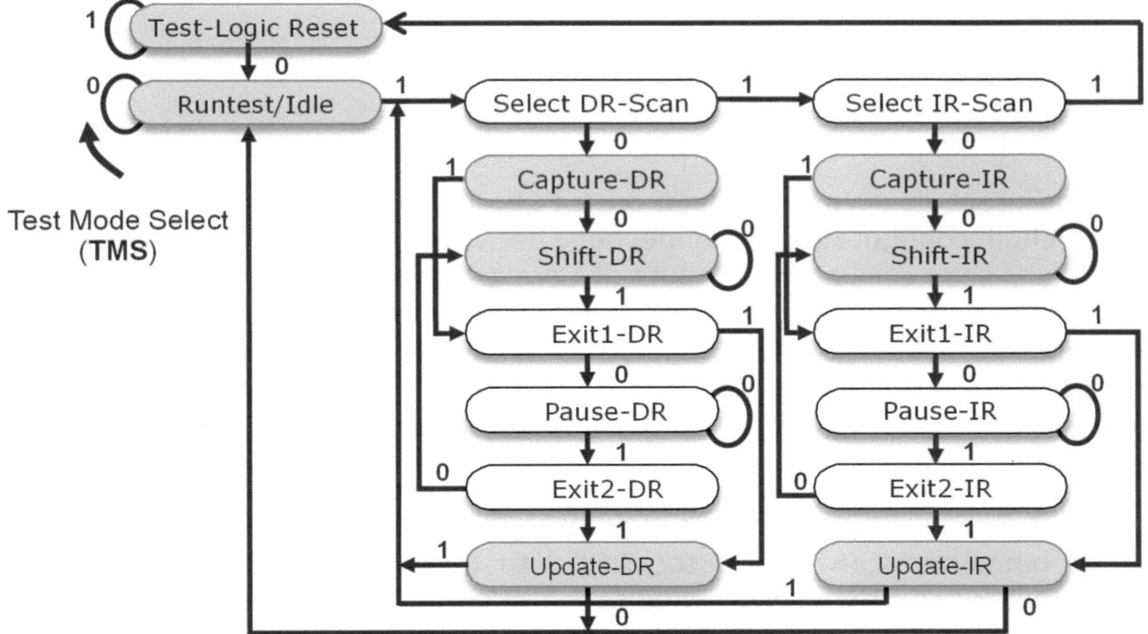

Figure 78: IEEE 1149.1 state diagram

DR'. These states trigger actions on controlling signals for the selected test register.

A typical application using boundary-scan logic on a PCB is shown in Figure 79.

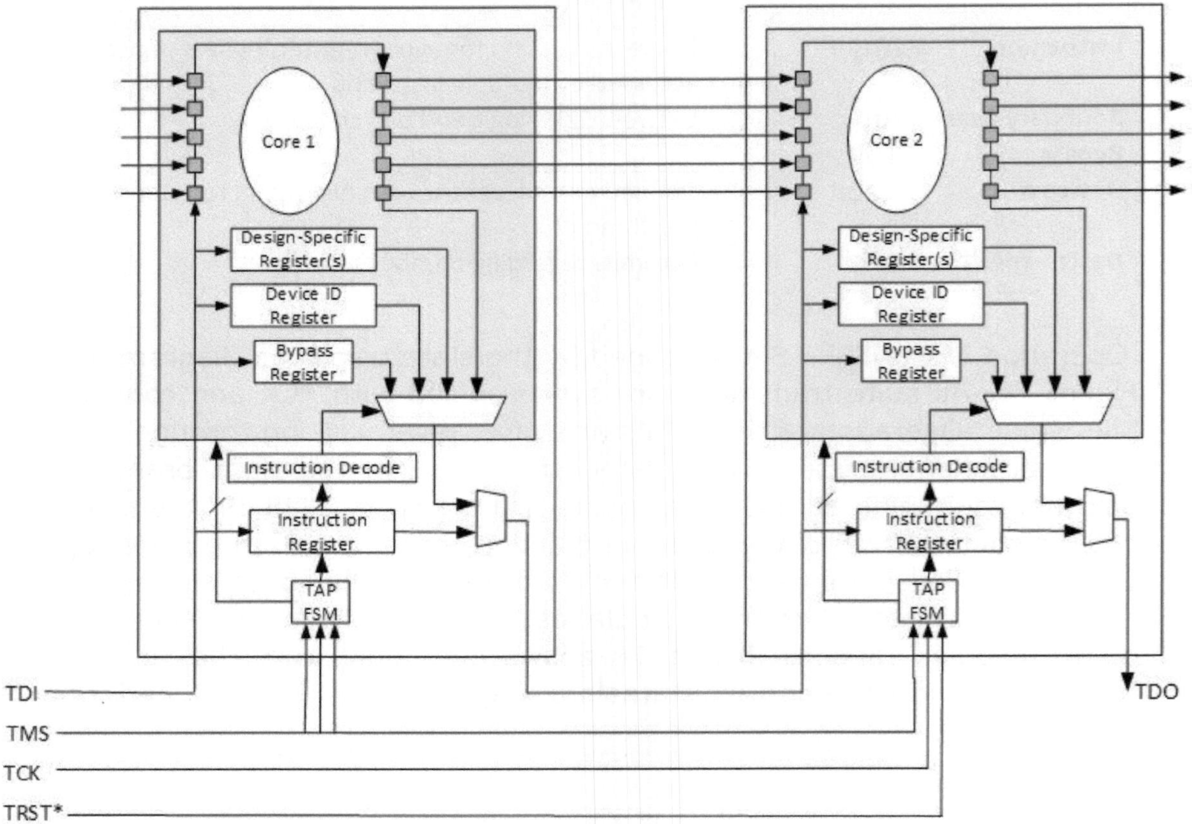

Figure 79: Using boundary-scan on PCB

Two different devices are placed on a PCB. TDI and TDO are connected in a daisy-chain configuration. If connections between Core 1 and Core 2 are tested, boundary-scan cells at the outputs of Core 1 are loaded and the results of the test are captured by the boundary-scan cells at the inputs of Core 2. This is an example for using the EXTEST mode as defined by the standard. EXTEST is a mandatory instruction required by the standard. Two more instructions are also mandatory, SAMPLE&PRELOAD and BYPASS. Here a short description of the mandatory instructions:

- EXTEST – Boundary-scan register is selected between TDI and TDO. The boundary-scan cells associated with outputs are preloaded with test values and those associated to inputs are put into capture mode.
- SAMPLE/PRELOAD – Boundary-scan register is selected between TDI and TDO while the device is still in functional mode. Data can be sampled into the boundary-scan register or pre-loaded into the register.

- BYPASS – Bypass register is selected between TDI and TDO. This register has a length of 1 bit and offers the opportunity to bypass the long boundary-scan register. Taking the daisy-chain example from Figure 79. Here for Core 1 device the bypass register is selected if only operations on boundary-scan register of Core 2 device are executed.

There are some more instructions defined by the standard but not a mandatory requirement. An example of such an optional instruction is IDCODE. This instruction is needed if a device ID register is included, and it defines the read-out of the product-specific 32bit identification. Another instruction very useful for manufacturing test is HIGHZ which allows to switch all tri-state output signals into the high-impedance state as soon as the instruction is valid.

Figure 80 shows the principle of a universal boundary-scan cell. Functional data are going from left to right. Serial test data are passed from bottom to top. Output 'ToTDO' is connected to 'FromTDI' of the next cell. The last cell of the boundary-scan register is followed by device's TDO pin as the TDI is connected to 'FromTDI' of the first register cell. Control and clock signals 'ShiftDR', 'Mode', 'ClockDR' and 'UpdateDR' are provided by the TAP controller. Their values depend on the state of the TAP-FSM and the selected instruction. A more practical realization of test register cells using TCK signal for clocking the two flip-flops in Figure 80 is discussed in Chapter Test control.

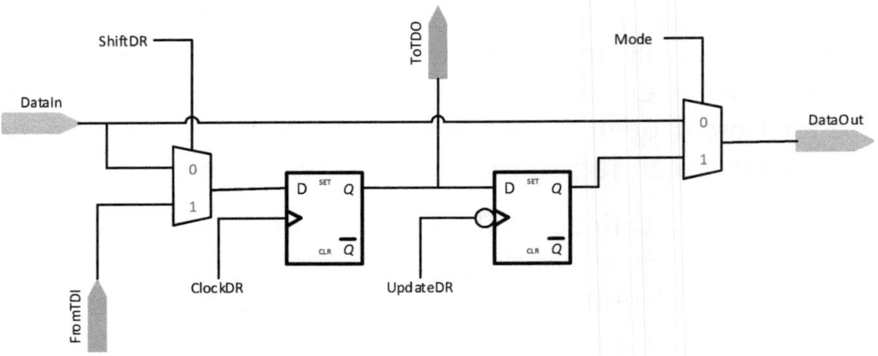

Figure 80: Universal boundary-scan cell (BC_1 type)

The shown cell can be used for input as well as for output pads. Figure 81 indicates the usage of boundary-cells for different I/O types. For an input pin a single cell (as shown with Figure 80) is used. A tri-state output pin needs one cell for the data path, another one for the output enable control signal. A bi-directional pin requires three boundary-scan cells.

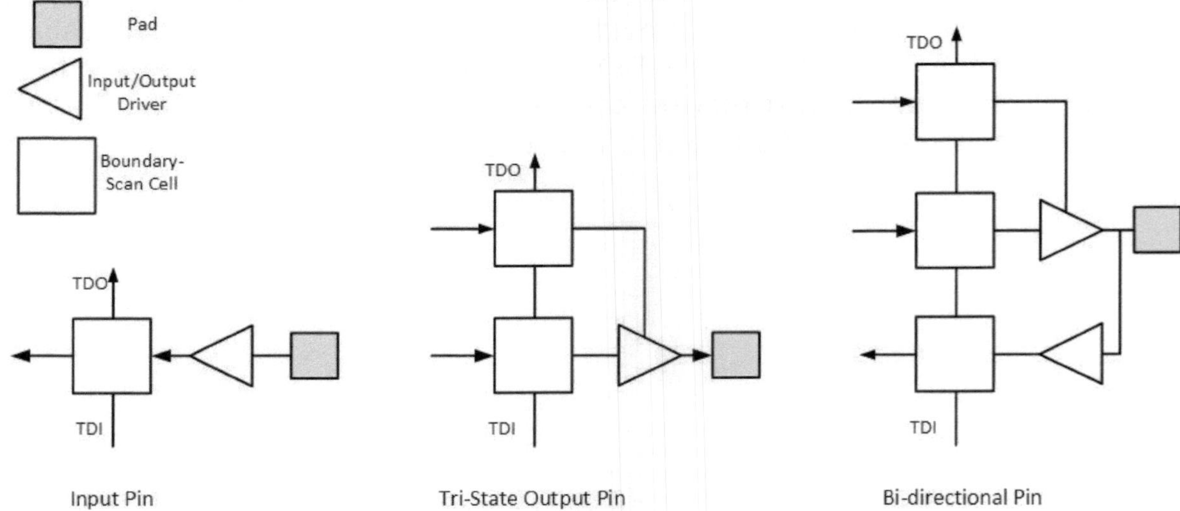

Figure 81: Boundary-scan cells for different I/O types

To describe the configuration of a device's boundary-scan system the Boundary-Scan Design Language (BSDL) format is defined. This format is human- and machine-readable. The complete definition is provided by IEEE 1149.1b standard [27]. The example below gives some elements of a BSDL description as pin mapping, instruction length and coding.

```
attribute PIN_MAP of XYZ : entity is
    PHYSICAL_PIN_MAP;

    constant DW:PIN_MAP_STRING:=

    "OE:1, Y:(2,3,4), A:(5,6,7), GND:8, VCC:9, "&
```

```
                      "TDO:10, TDI:11, TMS:12, TCK:13, NC:14";

attribute INSTRUCTION_LENGTH of XYZ : entity is 2;
attribute INSTRUCTION_OPCODE of XYZ : entity is
                                "BYPASS (11), "&
                                "EXTEST (00), "&
                                "SAMPLE (10) ";
attribute INSTRUCTION_CAPTURE of XYZ : entity is
                                "01";
```

Using BSDL allows software for PCB test pattern generation to get the required information about the boundary-scan systems for all devices. No information about the devices' mission functionality is required.

Based on the standardized architecture and BSDL information, specialized software tools not limited to PCB test but also for system debug and test purposes have established an ecosystem around IEEE 1149 standards. Companies providing software and services as their business offer tutorials and other information on their websites. Additional standards based on the original IEEE 1149.1 from 1990 have been released or are under development.

Test control

Based on the standardized protocol and the option to access an arbitrary number of test registers IEEE 1149.1 logic has been established as test control circuitry for test modes. These test modes enable the device to run scan test, memory test or any IP-specific test after manufacturing. For complex SoC many thousands of test register bits have to be configured to enable the execution of a dedicated manufacturing test pattern. A structured, hierarchical concept for test control logic is shown in Figure 82.

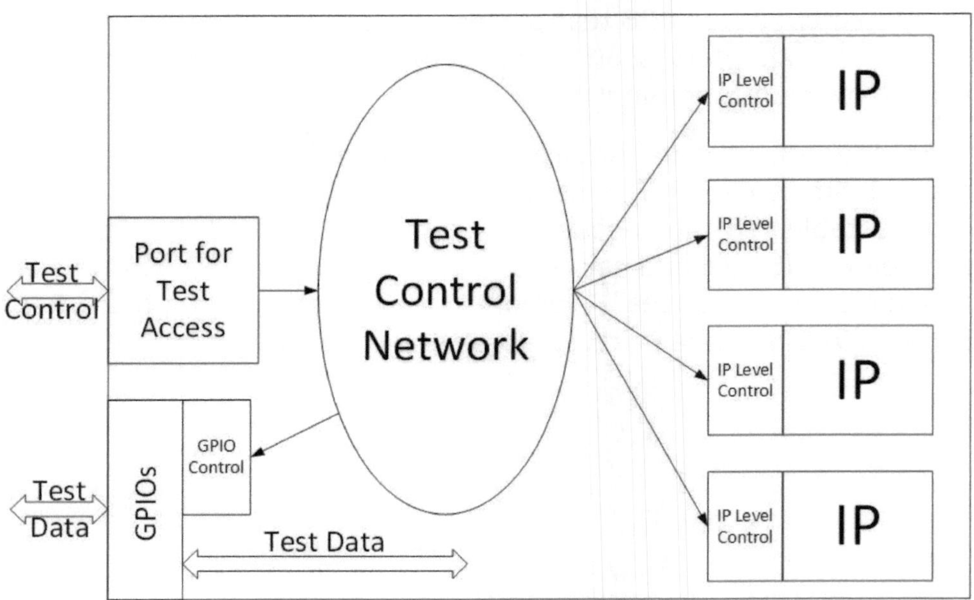

Figure 82: Hierarchical test control logic

The concept presented in Figure 82 uses a test control interface and a test data interface. The test control interface sets the device into different test modes. Here a low bandwidth interface is required, and the 5-pin TAP interface of IEEE 1149.1 is a perfect candidate. Using the TAP interface test registers compliant to this standard can be accessed to provide required settings for a dedicated test mode. Results from internal BIST can also be read out via this interface.

A second conceptually required interface in Figure 82 is for provision of test data. This handles e.g., scan data input and output which require a high bandwidth. Typically, General-Purpose Input-Output (GPIO) ports are used here. GPIO are normal functional pins which have multiplexing logic to enable test data functionality. Other test functions than scan use the GPIO to access internal nodes during a certain test mode to stimulate and observe these directly by the ATE. The GPIO interface is also controlled by the test control network.

IP level control blocks shown in Figure 82 are test registers. The number used for IP level test control and the length of the individual registers is not limited.

For a small number of test registers the test control network from Figure 82 could be realized by direct connections from TAP to the individual test registers. For more complex networks using hundreds of distributed test registers IEEE 1687 standard was developed.

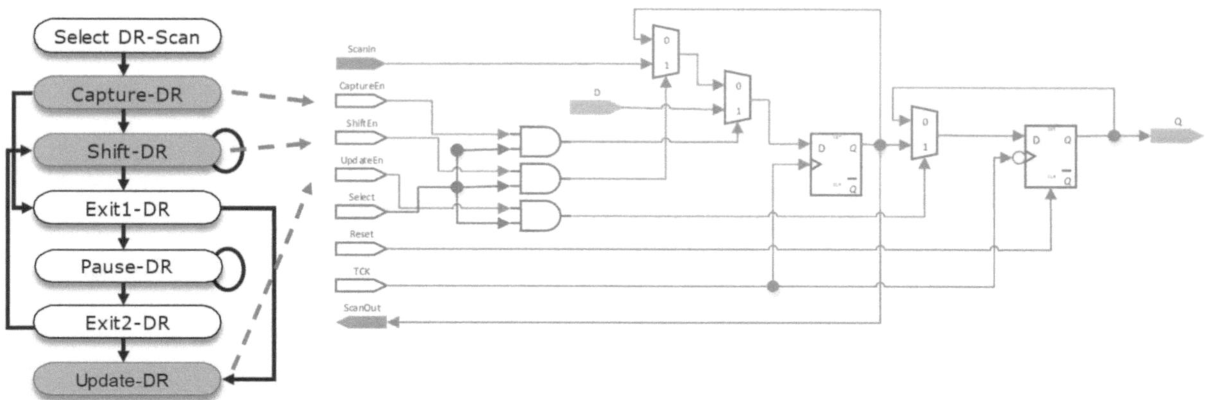

Figure 83: Test register and TAP-FSM states

Circuitry of a single bit test register cell is shown in Figure 83. Two flip-flops, multiplexers and AND-gates are included. The two flip-flops are controlled by TCK clock. The left one changes values at rising edge and is part of the serial stage starting at ScanIn, which is typically an element of a multi-bit test register. The serial stage ends at ScanOut which is either the output of a test register or is connected to ScanIn of the next register bit. The right flip-flop in Figure 83 changes value at the falling edge of TCK. This flip-flop represents the parallel output stage of the cell and is providing the value for parallel output Q. Depending on the TAP-FSM state exactly one of the enable signals CaptureEn, ShiftEn or UpdateEn can be active. If the register is selected, i.e. Select input signal is active, an operation is executed at the register. A capture operation captures data via D input into the serial stage flip-flop. During shift operation data is shifted from ScanIn into the serial stage flip-flop and to the ScanOut output. The update operation transfers data from the serial stage into the flip-flop of the parallel stage.

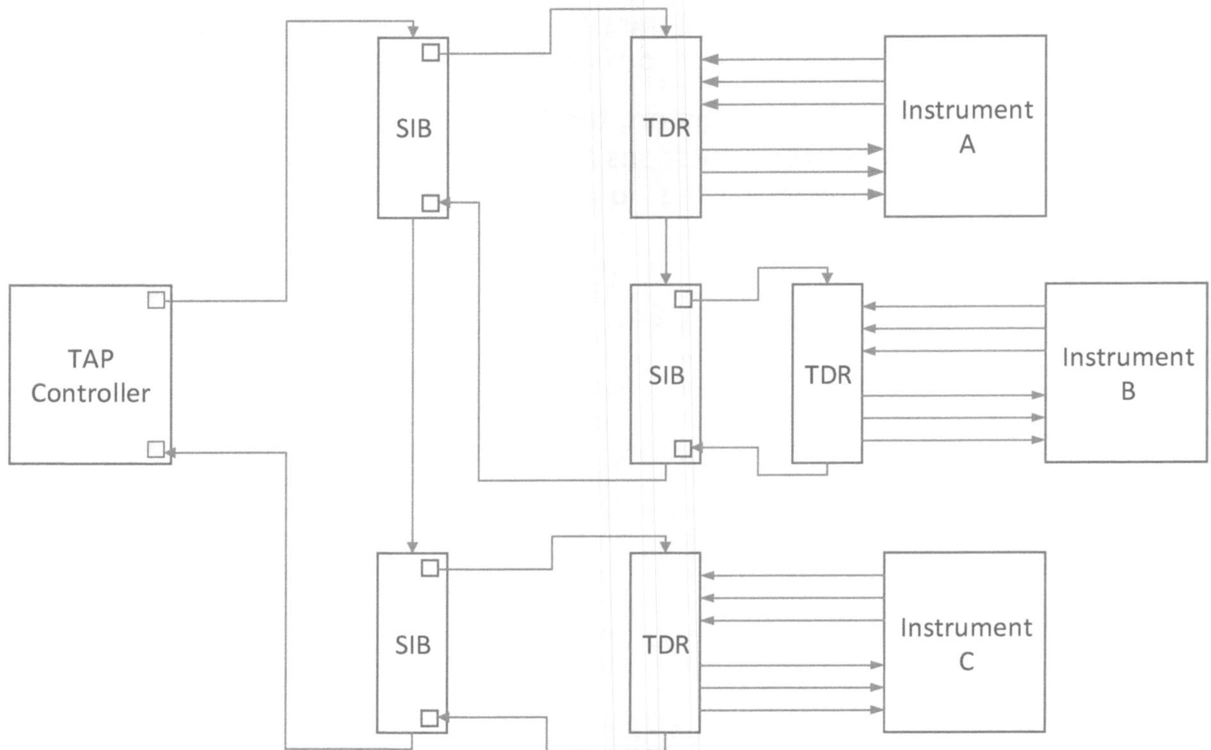

Figure 84: Example for iJTAG network according IEEE 1687

A test control network based on IEEE 1687 is shown in Figure 84. The network is connecting an IEEE 1149.1 compliant TAP-Controller with instruments distributed within a SoC. An instrument in the context of the iJTAG standard is any object which is tested, provides data or allows to be programmed via the TAP-Controller. Examples are analog mixed-signal IP, MBIST, scan test or any on-chip sensor used to control temperature or voltage. The network contains test data registers as explained above and shown in Figure 83. Additionally, Segment-Insertion-Bits (SIB) which are introduced by IEEE 1687, are used to allow reconfiguration of the test control network. A SIB provides access to a part of the network or set it into bypass mode to allow faster access to other TDR and instruments. The logic realization of a SIB is shown in Figure 85.

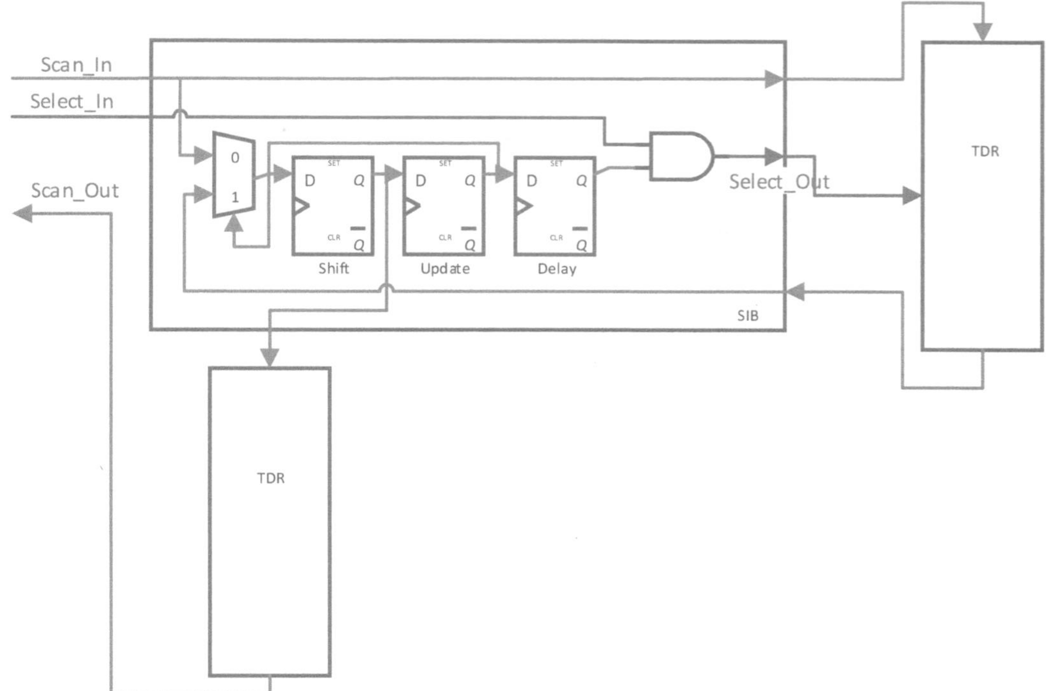

Figure 85: Realization of a Segment-Insertion-Bit (SIB)

The SIB cell uses 3 flip-flops for Shift, Update and Delay functions. Control signals for Shift and Update are identical to those shown in Figure 84. Synchronous clocking is also provided by TCK. The upper right TDR in Figure 85 is accessed only if select signal is active and the Delay flip-flop contains a logic '1'. Otherwise, the data is shifted via Shift flip-flop to the serial SIB output.

Using SIB cells allows the change of selected TDR without changing the TAP-Controller instruction. A single select signal is used for the complete network of TDR and SIB elements. Capture, shift and update signals are also identical for the whole network. The selection of accessible registers is controlled by the serial bit sequence. This flexibility allows concurrent test of various instruments. A TDR is selected and e.g., BIST is started. Within the next step another TDR is selected while the BIST is still running.

Power and Test

Power dissipation for CMOS integrated circuits has two components: Static and dynamic dissipation. The following formula shows the basic parameters:

$$\text{Power} = \alpha * f * C_L * VDD^2 + I_{Leak} * VDD$$

α: switching activity, C_L: lumped capacity; VDD supply voltage

Static, standby or leakage power goes linearly with supply voltage VDD, I_{Leak} increases with the number of transistors contributing to leakage.

Dynamic or active power is required to charge/discharge capacitive loads for connectivity lines and gate inputs. This component goes quadratic with VDD and linearly with frequency and switching activity.

For continuous increase of transistors per area power densities are also increasing. This results in two problems [29]:

- Supplying adequate power for circuit operation
- Heat flux from resulting dissipation

Both topics have influence on selected package technology.

During manufacturing test, the following power-related topics have to be considered:

- Power availability and power quality during test
- Higher switching activity during test
- Exponential increase of leakage current with higher temperatures
- Power management schemes disabled during test

Power availability and power quality during test

Reduced power availability during manufacturing test can impact the performance of the DUT and result in false fails, i.e., yield loss by test instabilities. Power supply can be impacted by longer connectors from ATE to DUT. Reduced pin count test which is typical for wafer sort insertion allows to connect a limited number of VDD and ground pins only. Current limiters on ATE to prevent burn-out are another source of low power availability and quality during test.

Higher switching activity during test

Structural tests as scan or memory BIST tend to have a higher activity than normal functional operation. The requirement for short test times leads to a compact scan test set. With a single scan vector multiple fault paths are activated at once. Memory BIST is a perfect example where as many memories as possible are tested concurrently. Power dissipation here is typically also above any functional use case. According to the formula above higher activity α during test can be compensated by reduced frequency f.

Low power scan pattern

ATPG programs have the capability to create patterns with reduced activity for shift and capture phases. Low power shift vectors are created by replacing don't care values by constant fill values. This results in a reduced activity during shift phase. With inactive low power mode random fill is used. With random fill more faults are typically detected with a single test vector than with constant fill. The resulting pattern without low power mode activated is typically smaller.

For low power capture mode, the ATPG tool makes use of on chip clock gating cells. An example for a clock-gating cell is shown in Figure 86. Here the algorithm gates clock domains which contain non-targeted faults for a specific vector.

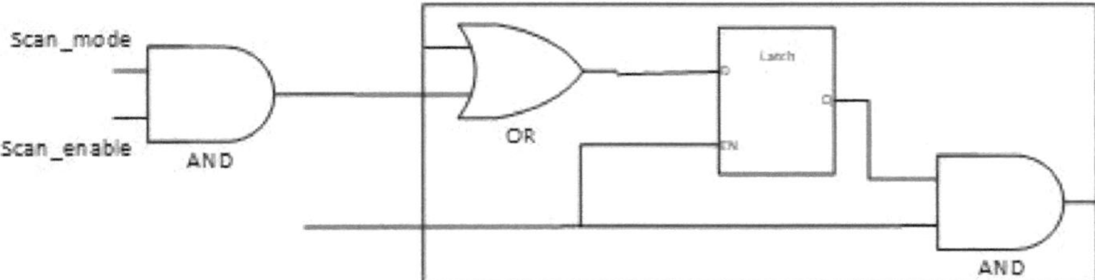

Figure 86: Clock-gating cell

Also, low power capture mode increases the overall pattern size. To achieve same fault coverages as power-unconstrained ATPG more vectors are needed. Consequently, test time is increased by using low power ATPG.

Exponential increase of leakage current with higher temperatures

As shown in Figure 87 active power dissipation is constant over temperature. Especially for high temperatures – and for high temperature test insertions – leakage power shows an exponential behavior. There is a risk of thermal runaway if temperatures are uncontrolled.

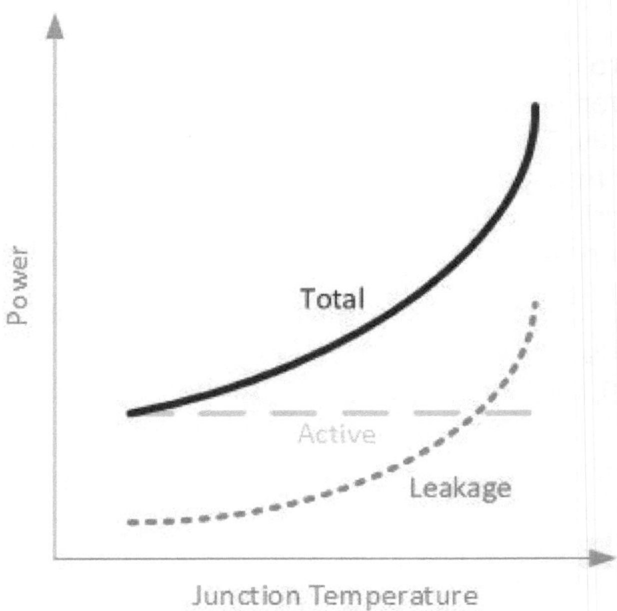

Figure 87: Exponential increase of leakage power for higher temperatures

Power management schemes disabled during test

To reduce power dissipation for functional operation various measures for static and dynamic power limitation have been introduced the last years. The most popular ones are power gating, power islands, dynamic frequency and voltage scaling. For any test execution it must be considered if a specific power savings measure has to be switched off. Since manufacturing test has to guarantee complete voltage and frequency ranges a disabling of power saving measures during test is for many cases impossible.

Security and DfT

An overview about DfT and test for security-critical products is given in [30]. This article provides a suitable structure for the topic and the following is mainly based on it.

Security covers concerns regarding data confidentiality and intellectual property protection. This can be protection of details on circuit's structures, architectures or even the Silicon manufacturing technology. Per definition (see Chapter DfT Tool Box) DfT increases the observability and controllability of circuit's internal nodes while security typically requires the opposite. The most obvious example are scan chains which can be used by an attacker to unload data during execution of security functions or to insert malicious data into the design.

For a security-critical circuit three criteria must hold for DfT techniques:

- Protected mode: Access to testing features restricted to privileged user
- Confidential information leakage: Confidential data handled by the circuit must never be output
- Malicious data insertion: In functional mode, the circuit should never process data inserted via the test interface

The most likely test-based attack scenario is after the device is deployed and an in-field hacker attempts to activate test features. Other possible attack scenarios are by test engineering during full access phases in test modes and implementing malicious IP provided by an external party. The last two scenarios are less likely. They are covered by development and manufacturing processes which are special to secure-critical products.

Counter measures can be classified into four categories:

- Protocol level
- Scan chain protection
- Communication security for test channel
- Pattern watermarking

Protocol level measures target to have a clear separation between test and user modes. The circuit is completely reset before entering test modes thus no data can be read out. Leaving test modes requires another reset to avoid that malicious data are brought into the product.

Scan chain protection covers measures like scan data scrambling, special protection for the scan enable signal or the insertion of a 'spy flip-flop' to detect any scan chain activation during user modes.

The security of the test channels covers measures to protect the on-chip test control network by malicious IP which is able to sniff, inject or modify bits in the test control network.

Pattern watermarking is a method to protect scan chains by comparing some of the scanned-in data. Using a golden reference for these data an authorized access can be checked.

Test and DfT for special structures

Testing specialized structures on integrated circuits requires advanced methodologies. These structures include High-Speed IO interfaces, Analog-Mixed Signal modules, and specialized memories such as DRAM and Flash.

For each of these structures, state-of-the-art testing techniques are employed to ensure optimal performance and reliability. When Design-for-Test (DfT) circuitry is available for these components, a detailed description of the DfT methods is also provided.

High-Speed IO Interfaces

High-Speed Input-Output (HSIO) interfaces have differential signal pairs for receive (Rx) and transmit (Tx) paths. Some interface protocols are standardized and used for wide-spread applications. Examples for this are Universal Serial Bus (USB), PCI-Express or LPDDRx memory interfaces. Typically, these interfaces operate in the GHz range. As shown in Figure 88 a HSIO interface consists of differential IO structures, a physical layer PHY and a digital controller. PHY and IO structure are high-speed analog-mixed signal circuitry and Controller part is digital, synthesizable logic. There is full-custom digital circuitry at controller-side of the PHY.

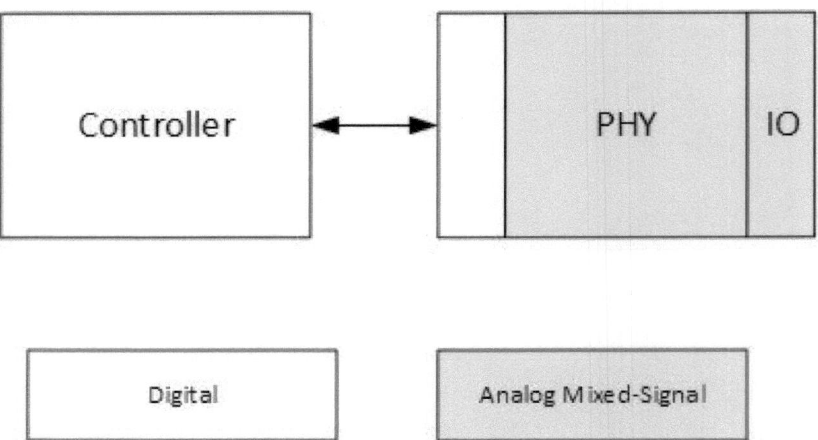

Figure 88: HSIO interface, general concept using controller and PHY part

A tutorial-like overview how to test such interfaces is provided in [13]. A summary is given in the following.

Protocol testing of HSIO interfaces would require costly ATE features. This includes pattern generation and data capture capabilities both at high-speed. Timing accuracy has to be in picosecond range and signal integrity needs to be well controlled. Such capabilities are still required for characterization test.

For high-volume manufacturing test a less expensive solution is needed. To allow use of low-cost ATE, loopback circuitry is established as DfT method for HSIO interfaces [14]. This circuitry includes on-chip pattern generation, pattern capture and comparison. After the completion of the typically required training sequence a re-start is required to execute the test.

Using loopback DfT, eye-diagrams can be constructed. Figure 89 (taken from [15]) shows an eye-diagram for a periodic signal. Eye-width (x-direction) allows the measurement of timing margins. Jitter is a measure for this defined as deviation of frequency and phase from the ideal signal behavior.

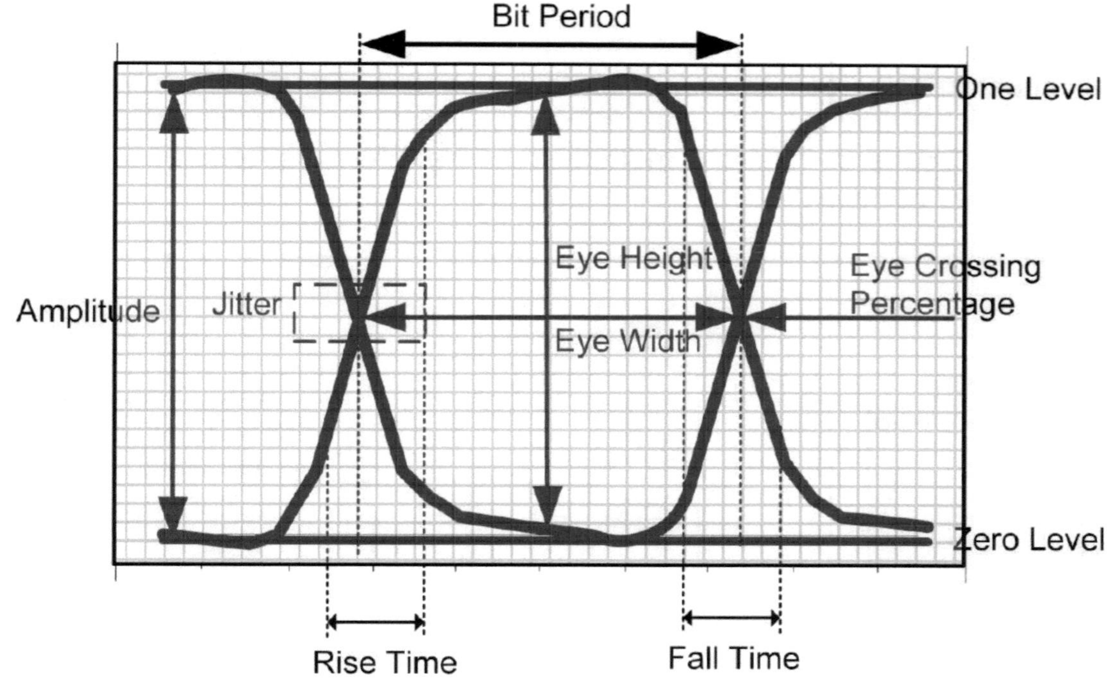

Figure 89: Eye-diagram measurements

Y-direction gives quality of signal levels or voltages of the periodic signal. Eye-height measurements allow evaluation of voltage margins.

Additionally, to the loopback-based tests described above tests for digital delay line and clock-data recovery blocks are required if available.

Table: Summary HSIO tests

Test type	DfT measure	Comment
Jitter	Loopback	Eye-diagram measurements
Clock-data recovery	Compare logic for phase interpolators	CDR not always available
Digital Delay Line	Ring oscillator mode	
Voltage margin	Loopback	Eye-diagram measurements

Phase-Locked Loops (PLL)

PLL are key elements for a SoC. Supplied with a reference clock higher clocks are generated to feed the on-chip clock domains. Figure 90 shows building blocks of an analog PLL with charge pumps. This figure is taken from [16]. The generated frequency F_{OSC} is N multiplied by the reference frequency PLLREF. N is an integer number.

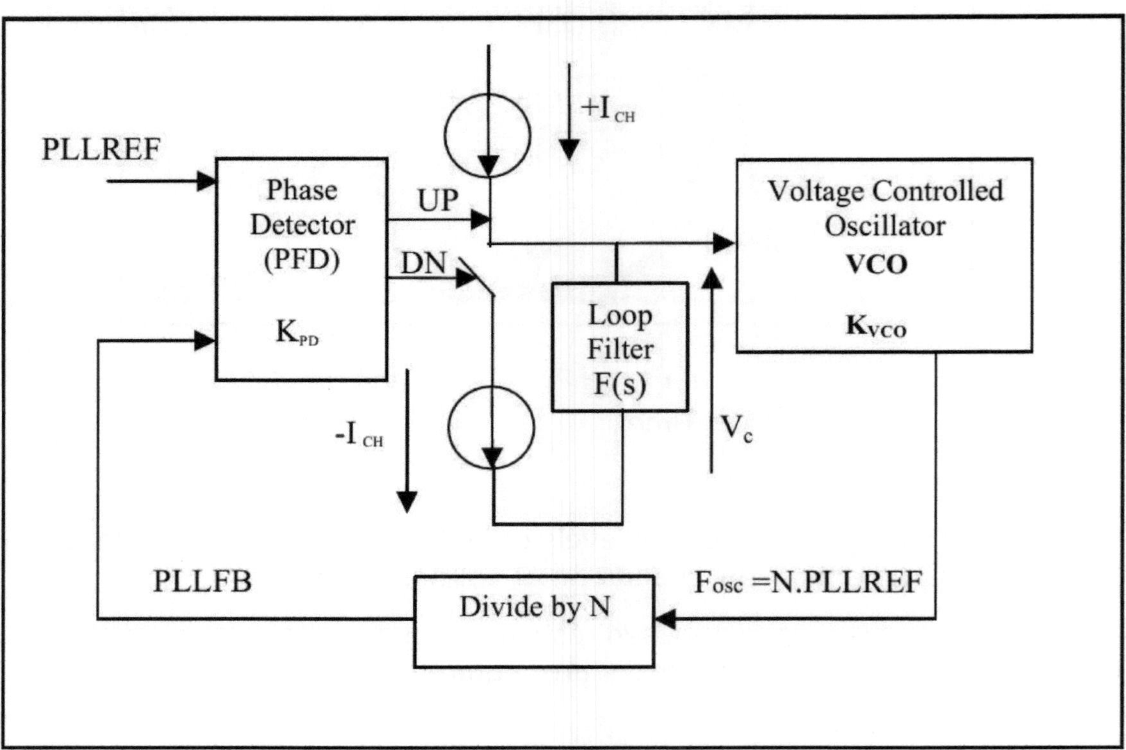

Figure 90: Analog PLL with charge pumps

Important for PLL is lock time T_{Lock}. This is the point in time when output frequency is stable and the chosen multiple of the reference frequency is achieved. Figure 91, also taken from [16], explains this important parameter. Typically, the following PLL parameters are checked during manufacturing test:

- Lock time
- Lock range (frequencies)
- Loop gain (achieved frequency can be calculated from here)
- RMS jitter

Using ATE to test these parameters directly would require high frequency and high precision capabilities. To avoid such costly equipment for high volume manufacturing BIST is very common for PLLs. In [17] a BIST solution allowing digital evaluation of the results is presented. Main elements of the BIST circuitry are a lock detector and a counter to evaluate the achieved frequency.

Also, an on-chip jitter measuring circuitry is proposed here. Jitter measurement is not always applied for high volume test since a high test time is required.

More recent designs are replacing PLLs with charge pumps according Figure 90 by all-digital PLLs. For these no charge pumps are required and digital filter

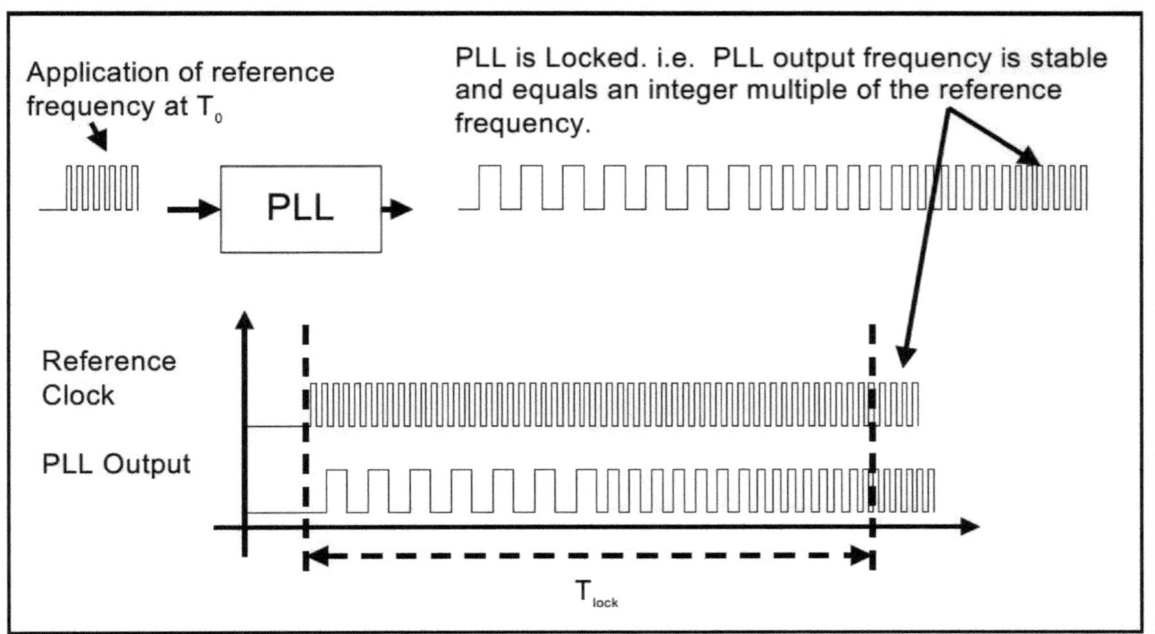

Figure 91: Lock time for a PLL

techniques are used. The VCO is replaced by a DCO [18]. The parameters to be tested are similar to those of analog PLLs:

- Frequency range
- PLL lock range
- DCO monotonicity
- Jitter

Also for all-digital PLL types BIST circuitry is used to enable efficient high volume manufacturing test.

Flash memories

Embedded flash arrays are non-volatile memory (NVM) types integrated into a SoC. They require additional mask layers compared to a standard CMOS technology.

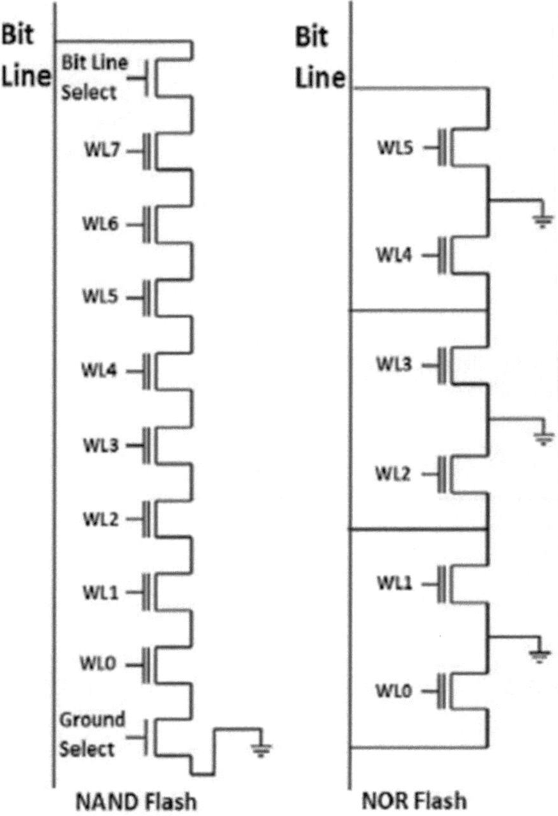

Figure 92: NAND and NOR flash types

Generally, NAND- and NOR-flash types are known (Figure 92). Both are based on the floating gate concept. Data are stored by changing the transistor threshold voltage V_{th}. V_{th} controls storing or removing charges on the floating gate. Flash memories support three main operations:

- Read
- Program
- Erase ('flash')

NAND-type flash has sequential read access while NOR-types have random read access as shown in Figure 92. Program and erase are faster for NAND than for NOR. No random access for the erase operation is available. Here, whole sectors or banks need to be erased. Generally, flash modules have a low power consumption and high cell density. During erase operation a high voltage which is enabled by charge pumps inject charges into the floating gate. For write or programming operation charges are removed from the floating gate using high voltage at memory's bit lines.

Even if embedded flash behavior is very technology-specific some common characteristics related to manufacturing test are known. To achieve acceptable yields redundancy and repair mechanisms are required. Some parameters need to be trimmed device specific. As part of the test flow wafer bake is essential. During this insertion step the whole unpowered and unconnected wafer is heated for a certain period. This baking step improves the bit error rate of the embedded flash arrays.

The nature of the floating gate technology leads to disturb, endurance and retention limitations. Disturb effects force neighbored cells to change their states during write, erase and read operations.

Typically, embedded flash tests are time consuming since erase and write are slow by nature. This limits the application of March-algorithms used for SRAM and DRAM tests. For an efficient test implementation, the design should be enabled for concurrent test of embedded flash arrays. Nevertheless, this may be limited by high current required for erase and write. Generally, tests for embedded flash need to be developed together with the providers of cell, technology and overall embedded flash architecture.

Test for embedded flash arrays needs to address

- Static and dynamic cell defects (including disturb)
- Data retention test (after bake insertion)
- Address decoding test
- Endurance test

As proposed in [19] and [20] test algorithms are using various background patterns during execution like solid, checkerboard and stripes.

DRAM

Dynamic Random-Access Memories (DRAM) incorporate the smallest memory cell. A single transistor and a capacitor enable storing of the information bit. This compactness allows high-capacity products. As a commodity product DRAM are used as main memory for PC and servers. Also embedded DRAM technologies are used. DRAM dies are used in SIP products to provide memory bandwidth for e.g. wireless modem products [21].

Using a capacitor as physical charge storage has the disadvantage that leakage current results in discharging of the capacitor. This is compensated by a refresh of stored data within regular intervals. Such an interval is called retention time. Checking this parameter is the most time-consuming element of DRAM test.

In [30] a typical DRAM test flow is described. As an important step to achieve acceptable yield redundancy activation is required for DRAM. A pre-fuse test identifies the defective cells. These are replaced by spare cells during a fusing step. A post-fuse test verifies the successful activation of the redundant elements.

After device assembly burn-in at high temperatures and elevated voltages reduces early fail rate.

To allow improvement of the manufacturing technology an option to provide bitfail maps is part of the test flow.

A discussion of test algorithm and stress conditions for DRAM test can be found in [22].

Test and DfT for System-in-Package Products

System-in-Package (SiP) products are semiconductor devices combining multiple dice inside a single package. For the customer it is not obvious if a device contains one or more Silicon dice. SiP products are used for the following reasons:

- Combining different Silicon technologies
- Provide smaller footprint devices by stacking of dice
- Overcome wafer yield limitations

The terminology used in scientific publications is 'Stacked integrated circuit' (SIC)[8]. Figure 93 (taken from [23]) gives examples for possible stack configurations.

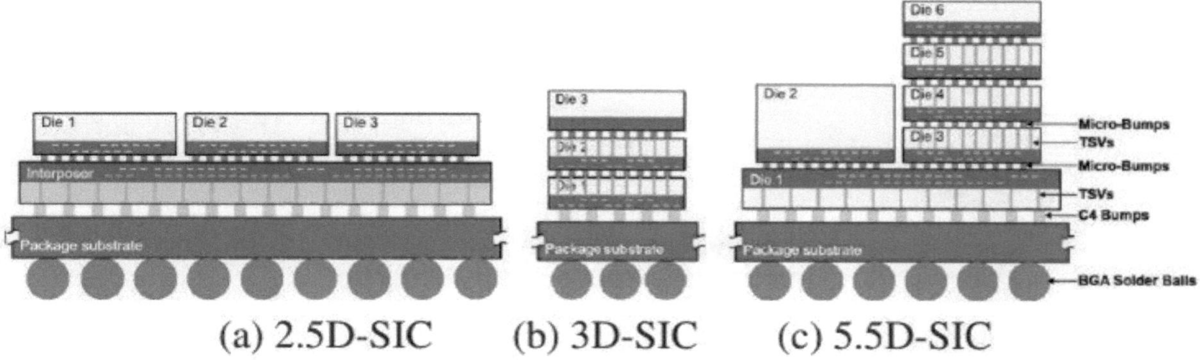

Figure 93: Types of Stacked integrated circuits

A special type of 3D-SIC stacking is a high-capacity Flash device used in USB sticks, portable devices or solid-state disk drives. Here identical dice are stacked to provide higher capacity while keeping device's footprint small.

Figure 93 also shows two interconnecting elements which are used in SIC devices only. Microbumps provide inter-die connectivity and have no direct access from the packaged device external interfaces. Through Silicon Vias (TSV) are implemented into the to-be-stacked dice providing top-down (or bottom-up) connectivity.

The test flow for SIC products is more cost sensitive than for single die devices. Wafer (or sort) test for all dice has to screen out all wafer-related yield loss. Identifying a defective die late in final test would result into screening out also the remaining dice inside a SIC. The cost assessment in Chapter General view on shift-left within manufacturing test flow has to add cost of all dice to the BE cost. A BE (or class) test for a SIC product could contain substrate, pre-bond, mid-bond and post-bond tests additional to the final test. Final test is required

---

[8] Term 'Chiplet' is used for huge area devices like server or graphics products. The terminology is not clearly defined here.

as quality gate. The other tests serve as cost optimization steps to identify packaging process-related yield loss earlier.

Generally, SIC products require special consideration for DfT architecture of the dice to be integrated. Let's consider the very common scenario of an SoC integrated together with a DRAM into an SIC. Communication between SoC and DRAM is established via an LPDDR interface. Connection lines between SoC and DRAM are not accessible from SIC package periphery. To enable test of the DRAM and the connectivity lines the LPDDR interface has to provide test routines during final test. These test routines are typically algorithmic similar to those used for BIST of embedded SRAM modules.

General DfT measures to support SIC products are standardized with IEEE 1838 [24]. A summary of this standard is given in [23].

From a conceptual perspective IEEE 1838 is seen as member of the IEEE 1149.x family. As shown in Figure 94 (taken from [23]) the first die at the

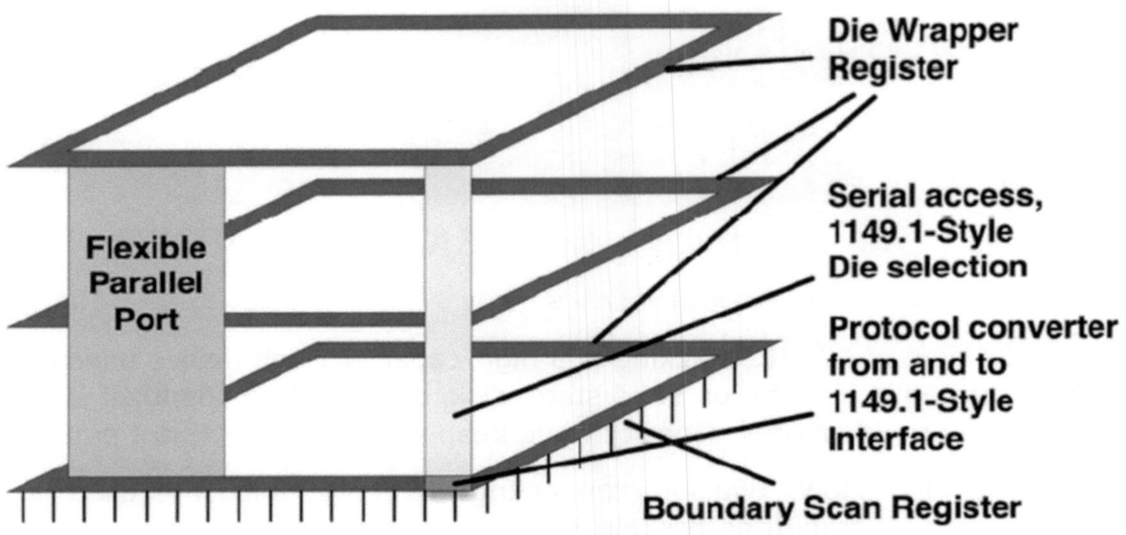

Fig. 2.   Elements of IEEE 1838

Figure 94: Elements of IEEE 1838

bottom contains a Boundary-Scan register and a TAP controller compliant to IEEE 1149.1. The first die inside a stack is defined as accessible from the product's boundary. The middle-die has connections to a predecessor- and a successor-die. The last-die is defined as one without any successor-die. IEEE 1838 defines

- A primary interface for middle- and last-dice
- A secondary interface for first- and middle-dice

Both interfaces are similar to the TAP-interface according to IEEE 1149.1. The primary interface connects to the secondary interface of the previous die. For middle- and last-dice Die Wrapper Register similar to IEEE 1149.1 boundary scan register are defined.

An optional Flexible Parallel Port is defined to enable high bandwidth test data. This port can be mapped to GPIO logic.

Test concept engineering

Test concept engineering plays a crucial role in identifying the optimal solution for dedicated products, aiming to achieve key business objectives such as quality, time-to-market, and cost-efficiency.

Manufacturing test concepts, along with associated Design-for-Test (DfT) strategies, work in tandem to optimize the complete testing system, both on-chip and at the Automated Test Equipment (ATE). The main driving forces behind these efforts are the prioritized business goals. This engineering process follows a meet-in-the-middle approach:

- DfT measures introduce the latest testing methodologies, pushing forward the test concept.

- Conversely, the costs associated with ATE testing influence the introduction of DfT features designed to shorten test durations or minimize the need for costly ATE capabilities.

An emerging term, 'DfX', encapsulates this multifaceted approach. Though not precisely defined, 'X' can represent various aspects such as:

- Test (i.e., manufacturing test)

- Reliability

- Validation (to support silicon validation)

- Characterization

- Diagnosis of low-yielding products or unstable test patterns (high data volume)

- Diagnosis of field returns (individual devices)

Focusing on manufacturing test concept engineering, its primary role is to ensure that products meet their specifications when delivered to customers. Therefore, the product specification is the foundational input for any test and DfT strategy. It outlines the product's components and global parameters, including power and performance. Nominal values with specified accuracies for all parameters must be verTest concept engineeringified against the capabilities of the target ATE.

Classification of DfT measures and test approaches

The description of DfT measures and test approaches in the previous chapters results into a characterization of tests into 4 classes:

Class	Characteristic	Typical structure or module	Test or DfT measure
A	Long test time w/o significant ATE resource usage ('Wait')	NVM, DRAM	High test parallelism
B	Minor ATE resource requirements	Memory BIST	Candidates for on-chip concurrency
C	Data intensive	Scan test	On-chip compression, Logic BIST
D	High measurement accuracy needed at ATE	Analog-mixed signal parameter tests and trimming	On-chip measurements and BIST concepts

- Class A tests are characterized by long test time without significant ATE resource usage. This is typical for testing DRAM and NVM products or embedded modules of these types.
- Minor ATE requirements exist for class B tests. Such tests require initialization, clock and power provision as well as test evaluation by ATE. All kind of BIST-based tests (e.g., for embedded memories) belong to this category. Loopback tests for HSIO modules belong also to this category.
- Class C tests are data intensive. Tests belonging to this class require ATE vector memory space which increases linearly with test length. All kind of scan tests are class C tests.
- For class D tests a high measurement accuracy is needed at ATE. This is required for most of the analog-mixed signal parameter tests. Such measurements are time consuming requiring advanced and costly ATE features. Trimming of internal circuitry is an additional step on top of the parameter measurements.

To achieve the business objectives a mix of DfT and test flow measures are used. A central aspect for optimization is the use of parallelism. To enable multi-site test for wafer and package test insertions the required bandwidth for test interfaces (e.g., for scan tests) needs to be considered. On-chip concurrent test has multiple facets and a key impact on the DfT architecture. The most obvious one is the parallel execution of BIST-enabled memory tests. The execution of all MBIST tests concurrently is limited by the availability of power only. Scan test is also limited by power. Without such a limit shift and capture for all internal scan chains could be done concurrently. Concurrent test in the sense that different pattern groups (e.g., AMS and MBIST) are executed in parallel needs some design and test problems to be solved. A more detailed discussion follows.

Optimizing costs by multi-site test

Whenever there are more ATE channels available than input/output structures on the DUT multi-site test is an option to reduce test cost. E.g., if ATE is equipped with 1000 channels and the device owns 200 balls 5 devices can be tested at the same time. For this example, no significant efforts are needed. Test time per device can be reduced by a factor of 5. Nevertheless, limits like power provision by ATE and mechanics like needle pressure have to be considered. Typically cost for ATE hardware (probe card, load board) increases for multi-site test. An example for testing four devices in parallel is shown in Figure 95. On the left side a single DUT is tested only and leaves a significant number of ATE channels unused. On the right side quad-site test is implemented increasing the usage of tester channels and increasing the throughput on ATE by a factor of 4.

Multi-site test can be further optimized by input broadcasting because identical devices are tested in parallel. Stimulation is typically identical, i.e., data sources from ATE can be connected in a broadcast mode to all devices tested in parallel. The same holds for clock and to some extent also for power channels. Inspecting again the 200 balls example above using data source and clock sharing between all test sites would allow 6, 7 or even more devices tested in parallel. The concept is shown in Figure 96.

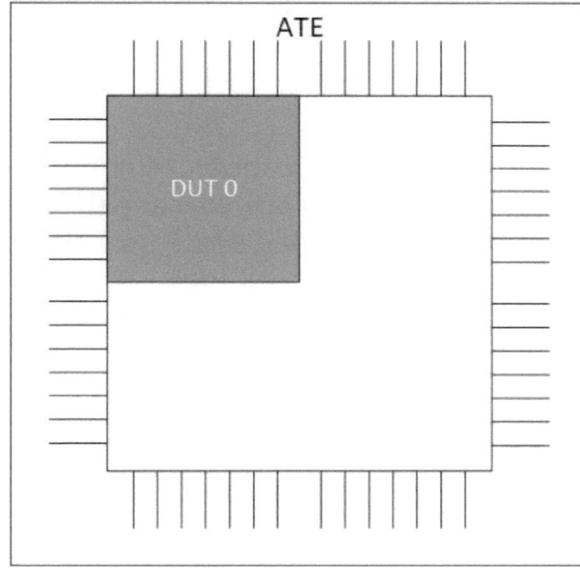

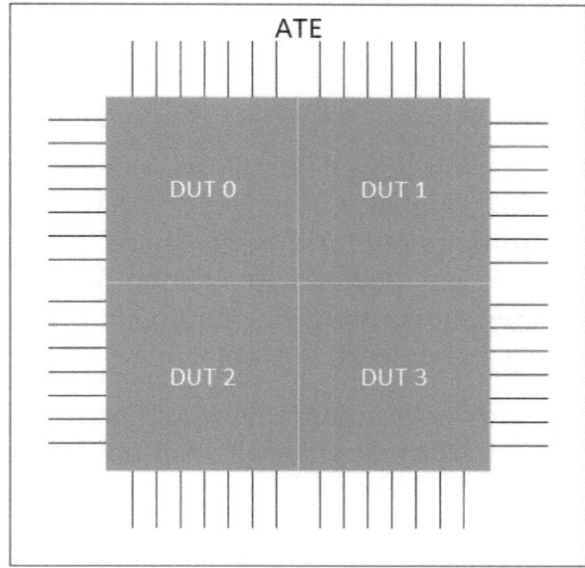

Figure 95: Effective use of ATEs tester channel by multi-site test (right part)

A multi-site approach very common for wafer test insertion is to not contact all pads. Only a sub-set is connected to ATE to run basic tests with a high overall coverage. The full test set is applied for device test with less devices tested in parallel.

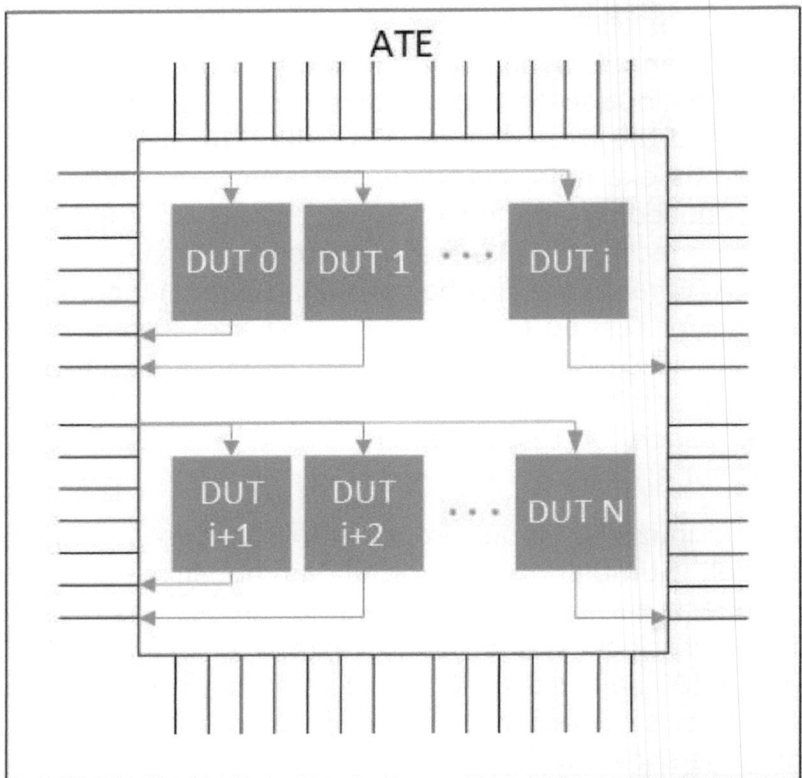

Figure 96: Multi-site test with input broadcasting

A high test parallelism is typically implemented for Class A type tests. This reduces the cost impact of long-running tests by the respective multi-site factor.

On-chip power domains are a key parameter for the specification of the multi-site capability. A power domain supplies a certain area with power independent from supply for others. There is also the option for embedded SRAM to have a different power supply than for the surrounding standard cell logic. A power domain contains dedicated supply pads associated with a certain voltage which again could be scalable.

Many power domains would allow high on-chip concurrency for e.g. during scan test. For tester parallelism any power domain needs a dedicated power supply by ATE. This reduces the multi-site capability for the test insertion.

A HSIO interface typically requires multiple power domains.

ATE and test flows

A typical ATE is configurable manufacturing equipment with has 1024 or more tester channels. For configuration cards are used providing e.g., 128 digital channels or a certain number of power supply channels. There are also specialized cards providing analog or RF stimulation and measurement capabilities available. For details please refer to ATE providers (e.g., https://www.advantest.com/products/soc/v93000.html, https://www.teradyne.com/products/ultraflex/). Generally, a pure digital ATE configuration is less expensive than one supporting analog or RF tests.

Especially Class A type tests require long test time without demanding expensive ATE resources. Often the device just has to 'wait' on the ATE for a significant portion of the test time. To overcome the issue of blocking expensive equipment by simple tasks here dedicated ATE is used. This is often combined by a massive multi-site test approach. A typical example is DRAM test where retention times of many seconds have to be confirmed during production test [30]. DRAM devices are high-volume commodity products and test patterns are algorithmic. Some parts of the ATE functionality can be skipped and by this ATE for DRAM becomes less expensive [32]. DRAM test is summarized in Chapter DRAM.

Another high-volume product category requiring long test times without high ATE interaction are Flash devices. Here erase and write cycles are dominating the manufacturing test time. A description of Flash test is described in Chapter Flash.

For a typical semiconductor product, a sequence of different test insertions is used. This sequence is also called test flow. A test insertion is characterized as follows:

- FE (sort, wafer level) or BE (class, device, package level) test insertion
- Temperature applied (hot, cold or room temperature)
- ATE and ATE configuration (test cell)
- Multi-site test factor

The test flow is defined to optimize manufacturing test cost for every shipped product by guaranteeing quality and reliability requirements. Consumer-like products typically have a test flow using room temperature insertions only. High quality markets like automotive require hot and cold test insertions applying temperatures according to the product's specification.

As an example, a test flow for flash memories is taken from [33] and shown in Figure 97. Here WP1 and WP2 are test insertions on wafer, FT1 and FT2 are final or package level test insertions. 'UV Erase' and 'Cycling Test' are specific for this kind of flash product. Burn-in is used to screen-out early lifetime failures.

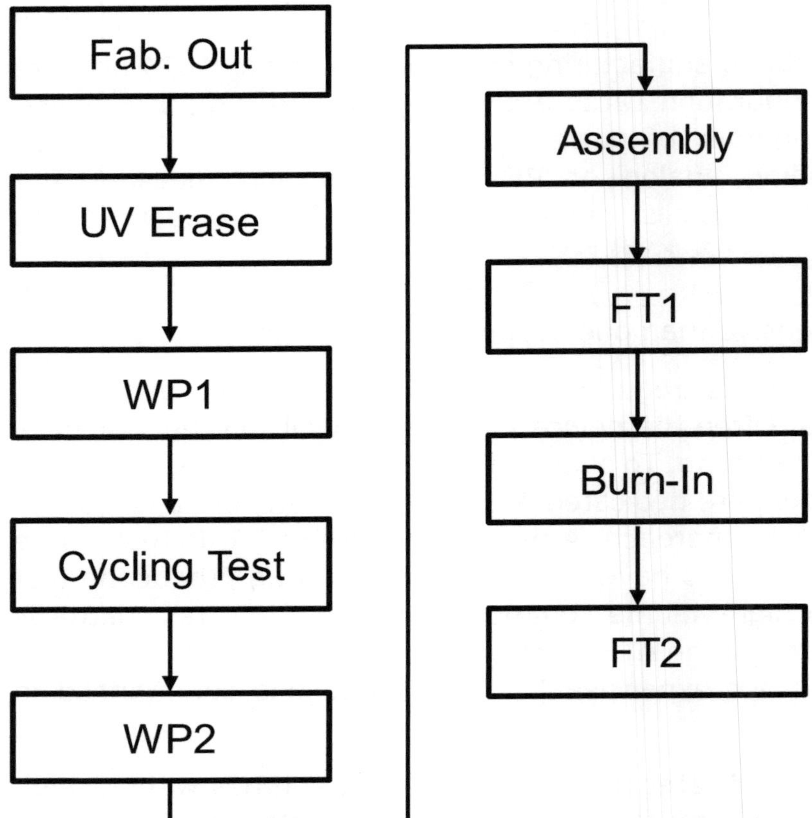

Figure 97: Test flow for flash products

The test content is different for all test insertions. Besides manufacturing test flows presented above there are also test insertions for non-high-volume activities like qualification, characterization, and validation tests. Here different equipment may be used, and test time optimization is not in focus.

A Systematic approach to test concept engineering

For testing a complex SoC divide-and-conquer approaches are used in the industry since years, see also [34]. Target of the test concept is to select the right test equipment, define the optimum test flow and have the right set of test patterns in place for every insertion. Main driving criteria are costs, quality, and reliability requirements.

Besides the identification of test equipment and the required test hardware, the test program running on the ATE is the key element for manufacturing test. A test program consists out of single tests ('patterns') which are typically grouped together ('pattern groups'). A pattern should be executable independent of any other test. This allows a reordering of tests within the test program for cost optimization if stop-on-first-fail approaches are used. In high volume manufacturing a simple pass-fail decision is required without additional diagnosis information. A stop-on-first-fail approach can be implemented for the first failing pattern group or for the first failing single pattern.

A test program contains a mix of IP-centric and device-centric patterns. A common separation into structural and functional patterns is used. The definitions are not always consistent. Generally, a structural pattern is not requiring any information about the function. E.g., a scan test is developed in the same way for a decoder, an adder, a CPU, or all other digital logic. The only required information is how the circuit is realized as gate level netlist ('structure'). A functional test requires detailed information about the function and is not looking for the structural realization. From application perspective a structural test does not require any processing unit or a complete boot of the device. Test control is in most cases done using test control registers accessible via JTAG interface. Functional tests are typically executed using an on-chip processing unit. For this a complete boot of the device is required and the test pattern is uploaded into the program memory. A typical example is the device performance test.

Here we describe a systematic approach for test concept engineering in cooperation with a product's DfT concept.

Step 1: IP-level test and DfT

Define test concept for every IP integrated, define required access to internal and external signals, overall guidance are quality requirements, identify suitable DfT measures either to enable IP-level test or reduce cost for test application

Step 2: Chip level tests

Define required chip-level tests like scan, MBIST and functional performance tests

Step 3: Business parameters

List of all tests required, including test time estimation and ATE resource requirements (pins to be contacted and channels required), also including expected yield loss

Step 4: Define test flow

Select target ATE, including features like number of channels, power supplies, timing performance and accuracy, analog measurement capability, vector memory depth and configuration options, define test cells with ATE, handler or prober and multi-site factors

General view on shift-left within manufacturing test flow

A manufacturing test flow has to guarantee that a product is manufactured as expected. This simple requirement is satisfied by a single test step at the end of the packaging or assembly process. In Figure 5 the last test insertion (package or BE test) would be the only one. A second requirement has to meet business targets to achieve minimum cost. For this FE or wafer test insertions are used. Identifying failing parts before assembly avoids the often costly assembly steps. Another reason for FE test is the identification of wafer-only yield. Feedback to wafer manufacturing on distribution of fails on the wafer (e.g., edge or 'donut' distributions) and the kind of fails is important information to control and optimize silicon manufacturing. Especially during yield learning phases wafer test is unavoidable.

The following thoughts consider the cost aspect only. At the end guidance is given to identify the right amount of FE test content. Parameters used are:

- Assembly cost (typically dominated by used package type, e.g. flip-chip vs. wire-bond): C_{Ass}
- BE test cost: C_{BEtest}
- Yield loss in BE: $Y_{loss} = 1 - Y_{actual}$
- FE test cost as product of ATE cost rate and test time: $C_{FEtest} = C_{ATE} * t_{FE}$

As mentioned above from a pure cost perspective FE test is introduced to save assembly and BE test costs. Thus, savings can be formulated as follows:

$$\text{Savings} = Y_{loss} * (C_{Ass} + C_{BEtest})$$

Y_{loss} covered in formula above represents only silicon wafer related yield loss, not assembly process induced fails. By introducing FE test insertion, cost for this should be lower than savings calculated above:

$$C_{FEtest} < Y_{loss} * (C_{Ass} + C_{BEtest}) \text{ or}$$

$$t_{FE} < Y_{loss} * (C_{Ass} + C_{BEtest}) / C_{ATE}$$

FE test time as a function of yield loss is shown in Figure 98 for three different cost rates.

Assuming assembly costs are the dominating ones for BE costs three different package types are considered. The upper line gives the most expensive package variant. Less costly examples are used for the lower lines with a less steep gradient. To realize savings by shifting a test associated with a certain yield loss from BE into FE the pair combining selected yield loss and FE test time needs to be below the line. E.g., to shift a yield loss of 3% in BE into FE by spending 8s FE test time would be a positive business case for the red FC line only. For the two less costly package types this would be above the lines and the business impact would be negative. Generally, long running tests with low yield loss should be kept in BE only. Short running tests with high yield loss potential should be executed in FE.

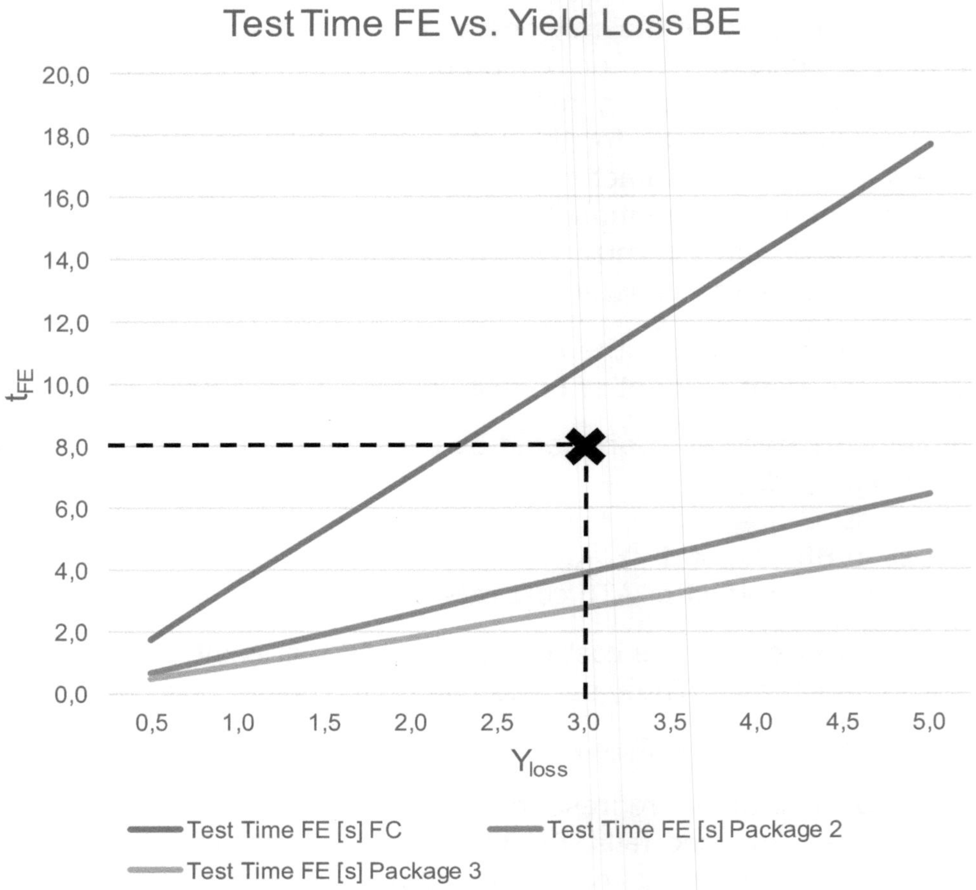

Figure 98: Test time FE versus yield loss BE

The principle shown above can be used as a shift-left model for test content and yield loss. Based on realistic assumptions for yield and test time criteria are available to select the right amount of FE test content. Looking closer into the example named 'Package 3' in Figure 98, which represents a very low cost packing technology, for high yielding products (i.e. low yield loss) FE test can be reduced drastically or even skipped completely.

Considering System-in-Package (SIP) products the gradient of the representing function is higher than any function shown in Figure 98. Assembly cost C_{Ass} increase drastically since two or more dice are integrated into a single package. For such products a high FE test time identifying almost complete wafer-induced yield is required for optimum CoS.

Table of terms

Test insertion	Defined by ATE used, pad or pins/balls contacted, temperature.
Sort insertion (also FE, or wafer test)	Needles are contacting pins ('touchdown'), typically multiple dice are contacted (multi-site test, see below), insertion for one wafer requires multiple touchdowns until all dice are tested.
Class insertions (also BE, class or device test)	Depending on package type balls or package pins are contacted, typically multiple dice tested in parallel (multi-site test, see below).
On-chip concurrency	Describes the number of elements for a certain structure tested in parallel. For scan test these elements are flip-flops, for embedded SRAM and ROM the number of instances tested in parallel.
Pattern concurrency	Is a new concept to run e.g., NVM and AMS test during the same time.
Multi-site test	Makes use of the physical properties (channels, power supply) of an ATE to test multiple dice or devices concurrently for a dedicated test insertion (e.g., sort at room temperature). Measure is the number of sites tested concurrently also called 'multi-site factor'.
Touchdown time	Gives the time needed to execute all tests of a sort test insertion for a group of dice tested in parallel. The sum of all touchdown times plus all index times (see below) gives the total wafer test time.
Effective test time	Used for calculating the test time per device for multi-site test $t = T/(N*Meff)$ N: number of dice or devices tested in parallel T: touchdown time for N dice (sort) or time to test N devices in parallel Meff: multi-site efficiency (≤ 1)
Index time	Time required to switch from one touchdown to the next or time required to switch from one group of devices to the next during ATE. This time is typically significantly lower than the effective test time.

References

1. John E. Price, "A new Look at Yield of Integrated Circuits", Proceedings IEEE, August 1970, pp. 1290-1291
2. ISO Standard 9000
3. T.W. Williams, N.C. Brown,"Defect Level as a Function of Fault Coverage", IEEE Transactions on Computers, vol. C-30, No. 12, Dec. 1981
4. O.H. Ibarra, S.K. Sahni, "Polynomially Complete Fault Detection Problems", IEEE Transactions on Computers, vol. C-24, no. 3, pp. 242-249, March 1975
5. P.H. Bardell and W.H. McAnney, "Parallel Pseudo-random Sequences for Built-In Test", International Test Conference 1984, pp. 302–308.
6. Paul H. Bardell, William H. McAnney, Jacob Savir, Built-In Test for VLSI: Pseudorandom Techniques, John Wiley & Sons, New York, 1987
7. Hortensius, P.D. McLeod, R.D. Pries, W.P. Miller, D.M. Card, H.C., "Cellular Automata-Based Pseudorandom Number Generators for Built-In Self-Test", IEEE Transactions on Computer-Aided Design, August 1989, vol.8, no.8, pp.842-859
8. Janusz Rajski, Nagesh Tamarapalli, Jerzy Tyszer, "Automated Synthesis of Large Phase Shifters for Built-In Self-Test", International Test Conference 1998, pp. 1047-1056
9. N. Mukherjee, D. Tille, M. Sapati, Y. Liu, J. Mayer, S. Milewski, E. Moghaddam, J. Rajski, J. Solecki, J. Tyszer, "Test Time and Area Optimized BIST Scheme for Automotive ICs", 2019 IEEE International Test Conference (ITC)
10. I. Hamzaoglu, J.H. Patel, "Reducing test application time for full-scan embedded cores", Proc. 29th Int. Symp. On Fault-Tolerant Computing (FTCS-29), June 1999
11. J. Rajski et al., "Embedded Deterministic Test", IEEE Transactions on Computer-Aided Design of Integrated Circuits and Systems, vol. 23, no.5, pp. 776-792, May 2004
12. O. Sinanoglu, A. Sehgal, E.J. Marinissen, J. Fitzgerald, J. Rearick, "Test Data Volume Comparison: Monolithic vs. Modular SoC Testing", IEEE Design & Test of Computers, May-Jun. 2009
13. S. Abdennadher, S. Shaik, "Practices in High-Speed IO Testing", 2016 IEEE European Test Symposium (ETS)
14. M. Jarwala, D. Le, Michael Heutmaker, "End-to-End Test Strategy for Wireless Systems",1995 IEEE International Test Conference (ITC)
15. On Semiconductor Application Note: https://www.onsemi.com/pub/Collateral/AND9075-D.PDF, "Understanding Data Eye Diagram Methodology for Analyzing High Speed Digital Signals", [downloaded January 19, 2021]
16. M. Burbidge, A. Richardson, Phase Locked Loop Test Methodologies, 2008, [downloaded March 15, 2021]

17. S. Sunter, A. Roy, "BIST for Phase-Locked Loops in Digital Applications", 1999 IEEE International Test Conference, 1999

18. R. Staszewski, I. Bashir, O. Eliezer, "RF Built-in Self Test of a Wireless Transmitter", IEEE Transactions on Circuits and Systems, vol. 54, no. 2, Feb. 2007

19. S. Martirosyan, G. Harutyunyan, S. Shoukourian and Y. Zorian, "An efficient testing methodology for embedded flash memories", 2017 IEEE East-West Design & Test Symposium (EWDTS), 2017

20. O. Ginez, J.-M. Daga, P. Girard, C. Landrault, S. Pravossoudovitch, A. Virazel, "Embedded Flash Testing: Overview and Perspectives", DTIS: Design and Technology of Integrated Systems in Nanoscale Era, 2006, pp.210-215

21. https://www.qualcomm.com/products/snapdragon-system-package [downloaded August 13, 2021]

22. A.J. van de Goor, "Industrial Evolution of DRAM Tests", IEEE Design & Test of Computers, Sept-Oct. 2004

23. E.J. Marinissen, T. McLaurin, H. Jiao, "IEEE P1838: DfT Standard-under-Development for 2.5D-, 3D-, and 5.5D-SICs", 2016 IEEE European Test Symposium (ETS), 2016

24. - , IEEE Standard for Test Access Architecture for Three-Dimensional Stacked Integrated Circuits, IEEE Std 1838-2019, 2019

25. S. Hamdioui, G. Gaydadjiev, A.J. van de Goor, "The State-of-art and Future Trends in Testing Embedded Memories", 2004 International Workshop on Memory Technology, Design and Testing (MTDT)

26. - , IEEE Standard for Test Access Port and Boundary-Scan Architecture, IEEE Std 1149.1-2013 (Revision of IEEE Std 1149.1-2001), 2013

27. - , IEEE Supplement to Standard Test Access Port and Boundary-Scan Architecture (1149.1), IEEE Std 1149.1b-1994, 1994

28. -, IEEE Standard for Access and Control of Instrumentation Embedded within a Semiconductor Device, IEEE Std 1687-2014, 2014

29. P. Girard, N. Nicolici, X. Wen (Editors), Power-Aware Testing and Test Strategies for Low Power Devices, Springer, 2010, pp. 31-64

30. D. Hély, K. Rosenfeld, R. Karri, "Security challenges during VLSI test", 2011 IEEE 9th International New Circuits and systems conference

31. J.E. Vollrath, "Testing and Characterization of SDRAMs", IEEE Design & Test of Computers, Jan.-Feb. 2003

32. K. Nakamae, H. Ikeda, H. Fujioka, "Evaluation of Final Test Process in 64-Mbit DRAM Manufacturing System through Simulation Analysis", 2003 IEEE/SEMI Advanced Manufacturing Conference

33. C.T. Huang, J.C. Yeh, Y.Y. Shih, R.F. Huang, C.T. Wu, "On Test and Diagnostics of Flash Memories", Asian Test Symposium (ATS), 2004

34. F.P.M. Beenker, R.G. Bennetts, A.P. Thijssen, Testability Concepts for Digital ICs - The Macro Test Approach, Kluwer Academic Publishers, 1995

35. https://en.wikipedia.org/wiki/Standard_cell [downloaded April 20, 2021]

36. J. Alt, U. Mahlstedt, "Simulation of Non-Classical Faults on the Gate Level - Fault Modeling", 1993 IEEE VLSI Test Symposium, 1993

37. F. Hapke, R. Krenz-Baath, A. Glowatz, J. Schloeffel, H. Hashempour, S. Eichenberger, C. Hora, D. Adolfsson, "Defect-Oriented Cell-Aware ATPG and Fault Simulation for Industrial Cell Libraries and Designs", 2009 International Test Conference, 2009

38. W. Howell, F. Hapke, E. Brazil, S. Venkataraman, E. Datta, A. Glowatz, W. Redemund, J. Schmerberg, A. Fast, J. Rajski, "DPPM Reduction Methods and New Defect Oriented Test Methods Applied to Advanced FinFET Technologies", 2018 International Test Conference (ITC), 2018

39. D. Kraak, M. Taouil, I. Agbo, S. Hamdioui, P. Weckx, S. Cosemans, F. Catthoor, "Parametric and Functional Degradation Analysis of Complete 14-nm FinFET SRAM", IEEE Transactions on Very Large Scale Integration (VLSI) Systems, vol. 27, No. 6, 2019

40. E.F. Moore, "Gedankenexperiments on Sequential Machines", Automata Studies, Annals of Mathematical Studies. Princeton, N.J.: Princeton University Press (34): 129–153, 1956

41. JP Roth, "Diagnosis of automata failures: A calculus and a method", IBM Journal of Research and Development, 10(4): pp.278-291, 1966

42. Z. Navabi, _Digital System Test and Testable Design_, Springer Verlag, 2011

43. S. Kaeshammer, _Configurable Test Access Mechanisms utilizing IEEE 1687 (iJTAG) compliant Network Structures_, Master Thesis FAU Erlangen, 2022

44. I. Koren, C. M. Krishna, _Fault-Tolerant Systems_, Elsevier, Inc., 2007

45. W. Chen, J. Bhadra, "Practices and Challenges for Achieving Functional Safety of Modern Automotive SoCs: A Tutorial Introduction", IEEE Design and Test, April 2019

46. https://www.ablic.com/en/semicon/products/automotive/asil [downloaded March 04, 2025]

Abbreviations

ADC	Analog-to-Digital Converter	
ASP	Average Selling Price	
ATE	Automated Test Equipment	
ATPG	Automated Test Pattern Generation	
BIST	Built-In Self-Test	
BITE	Built-In Test Equipment	Used for A3G, IFX internal term
CMOS	Complementary Metal-Oxide Semiconductor	
CoS	Cost-of-Sales	Also: manufacturing costs
CUT	Circuit-under-Test	Also: DUT
DFM	Design-for-Manufacturability	Also: Design-for-Manufacturing
DFT	Design-for-Test(ability)	
DPPM	Defective-parts-per-million	Dpm or ppm also used
DRAM	Dynamic Random Access Memory	
DUT	Device-under-Test	Also: CUT
ECO	Engineering change order	https://en.wikipedia.org/wiki/Engineering_change_order
EDA	Electronic Design Automation	
EDT	Embedded Deterministic Test	
FF	Flip-flip	
FSM	Finite-State Machine	
GPIO	General-Purpose Input-Output	

ICL	Instrument Connectivity Language	Defined by IEEE 1687
JTAG	Joint test action group	Synonymous to TAP-Controller
LBIST	Logic Built-In Self-Test	
LFSR	Linear-Feedback Shift Register	
LSB	Least-significant bis	
HSIO	High-Speed Input-Output	
MISR	Multiple-Input Shift Register	
MSB	Most-significant bit	
NVM	Non-volatile memory	
OCC	On-chip clock control	Used for pulse generation during dynamic scan test
PCB	Printed-circuit board	
PDL	Procedural Description Language	Defined by IEEE 1687
PLL	Phase-locked loop	
PRPG	Pseudo-random pattern generator	
RRAM	Resistive Random Access Memory	ReRAM also used as abbreviation
ROM	Read-Only Memory	
Rx	Receive	
SDF	Standard Delay Format	
SIB	Segment-Insertion-Bit	Standardized by IEEE 1687
SIC	Stacked integrated circuit	Used for SIP
SIP	System-in-Package	
SOC	System-on-Chip	
SRAM	Static Random Access Memory	
TAP	Test access port	

TCL	Tool Command Language	
TDR	Test data register	Also: Test control register
TSV	Through Silicon Via	
Tx	Transmit	
USB	Universal Serial Bus	
WLBI	Wafer Level Burn-In	

Appendix

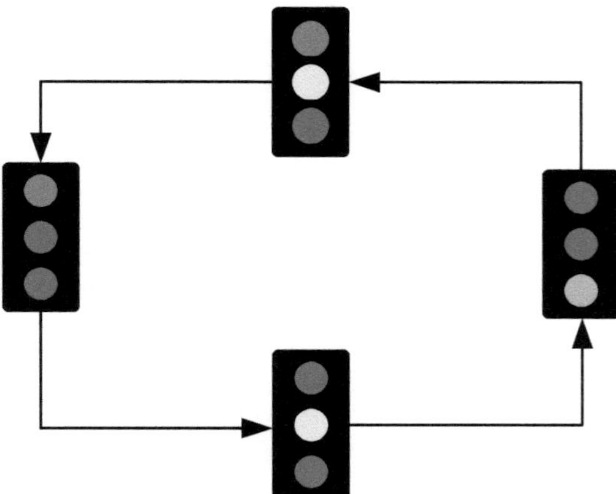

Example: (German) traffic lights have four different states

To store these states in registers log2 N = 2 state variables necessary

2 bits required to store information

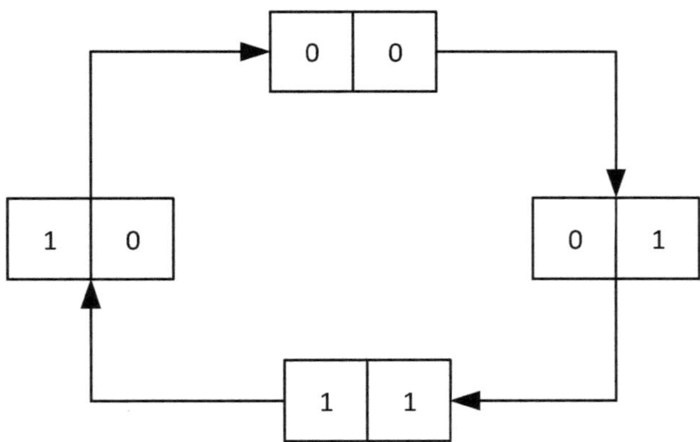

Here a <u>possible</u> encoding for traffic light example is shown.

Consider exponential behavior of state variables. Number of state increase with 2^r, r is number of state variables (or registers).

Boolean Logic – 2-valued

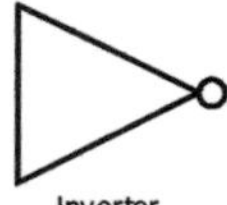

Inverter

A	INV
0	1
1	0

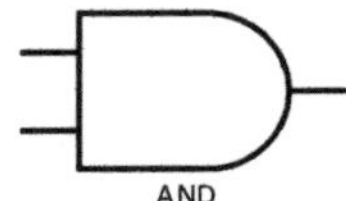 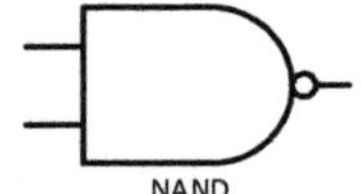

AND   NAND

A	B	AND	NAND
0	0	0	1
0	1	0	1
1	0	0	1
1	1	1	0

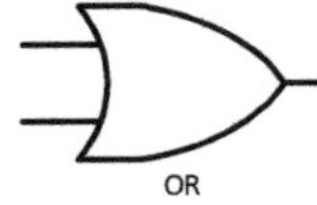 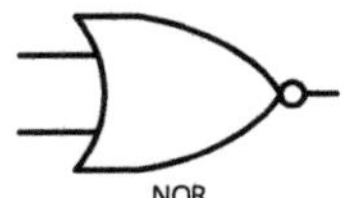

OR   NOR

A	B	OR	NOR
0	0	0	1
0	1	1	0
1	0	1	0
1	1	1	0

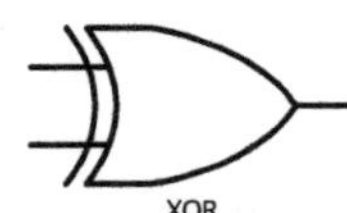 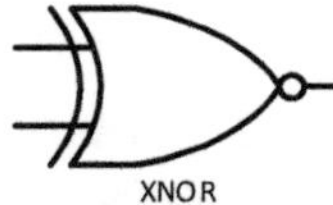

XOR   XNOR

A	B	EXOR	EXNOR
0	0	0	1
0	1	1	0
1	0	1	0
1	1	0	1

Boolean Logic – 5-valued

As used for D-propagation: Values are (0, 1, X, D, $\bar{D}$)

Tables not complete, take 2-valued logic tables additionally

A	INV
D	$\bar{D}$
$\bar{D}$	D

A	INV
X	X

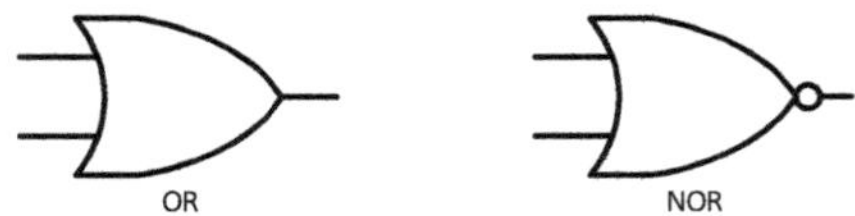

A	B	AND	NAND
0	D	0	1
D	1	D	$\bar{D}$
$\bar{D}$	0	0	1
1	$\bar{D}$	$\bar{D}$	D

A	B	AND	NAND
0	X	0	1
X	1	X	X
X	0	0	1
1	X	X	X

A	B	OR	NOR
0	D	D	$\bar{D}$
D	1	1	0
$\bar{D}$	0	$\bar{D}$	D
1	$\bar{D}$	1	0

A	B	OR	NOR
0	X	X	X
X	1	1	0
X	0	X	X
1	X	1	0

A	B	EXOR	EXNOR
0	D	D	$\bar{D}$
D	1	$\bar{D}$	D
D	0	D	$\bar{D}$
1	D	$\bar{D}$	D

A	B	EXOR	EXNOR
0	X	X	X
X	1	X	X
X	0	X	X
1	X	X	X

Complexity

The term complexity is used within different context types in our daily life, e.g., for human interaction, for problems of societies etc. General statement is, something is getting more complex if more elements are involved. Also, specialization increases complexity. E.g., if a house is constructed by more diverse specialists than a small number, more contributors have to be coordinated.

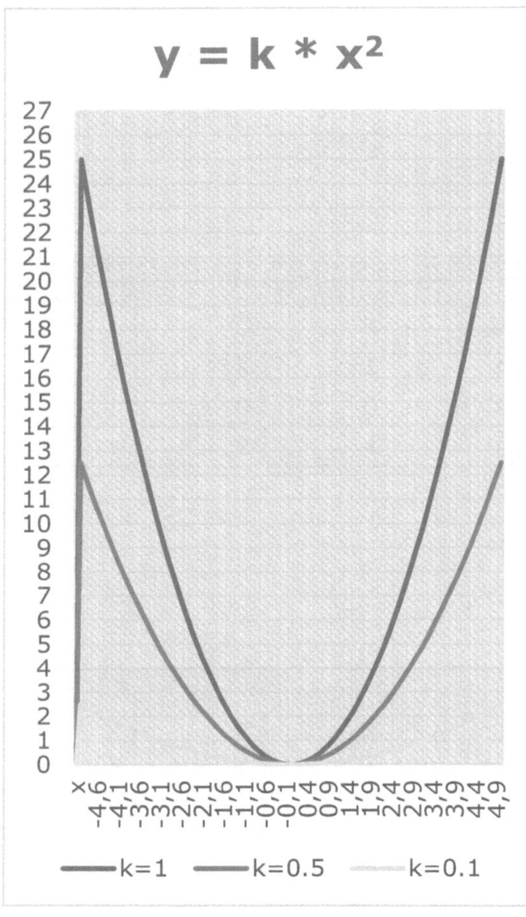

Complexity theory as used in engineering and computer science classifies solvable problems into complexity classes like linear, exponential, polynomial or other categories. An algorithm to solve a problem requires for execution a certain time and a certain memory (time and space complexity). In the graph shown examples are given of quadratic complexity. Values x represent the elements of the problem, y stands for the resulting effort. The constant k is independent of the problem size. As an example, k can represent the time to process one element. Using a faster processor for calculation k is reduced accordingly.

For complexity considerations the behavior of the algorithm for an increased number of elements is of interest. This is independent of k and for quadratic algorithms a doubling of elements increases the effort by a factor of 4.

Thus, constants are typically not considered to classify an algorithm. The following O() notation is used to describe behavior of algorithms:

Linear	O(n),	n: number of elements
Quadratic	$O(n^2)$	
Polynomial	$O(n^k)$	k > 1
Exponential	$O(d^n)$,	d > 1

Profit and Loss Statements are used on product level, for a business division or on company level aggregated in quarterly/yearly public reports.

Revenue

This is the sum of all money you get by selling your product(s), service(s) and other earnings. For products it's influenced by sales price (average selling price, ASP) and the volume you shipped to your customers. You also find sales as a synonym for revenue.

Cost types

General and administration (G&A): Administration costs including salaries for the executive board.

Sales: Costs for sales and marketing activities.

Research and Development (R&D): These are one-time costs and independent of the shipped volume. This cost type covers all product development efforts as well as product-independent research efforts.

Cost of Sales (CoS): This covers all costs to manufacture and to ship products. These are volume dependent, if there is no volume there are no cost of sales.

Profit

Profit is Revenue minus all types of costs. You need profit to pay dividend to company shareholders and to allow investments into new products, businesses or manufacturing facilities.

Volume-dependent optimization

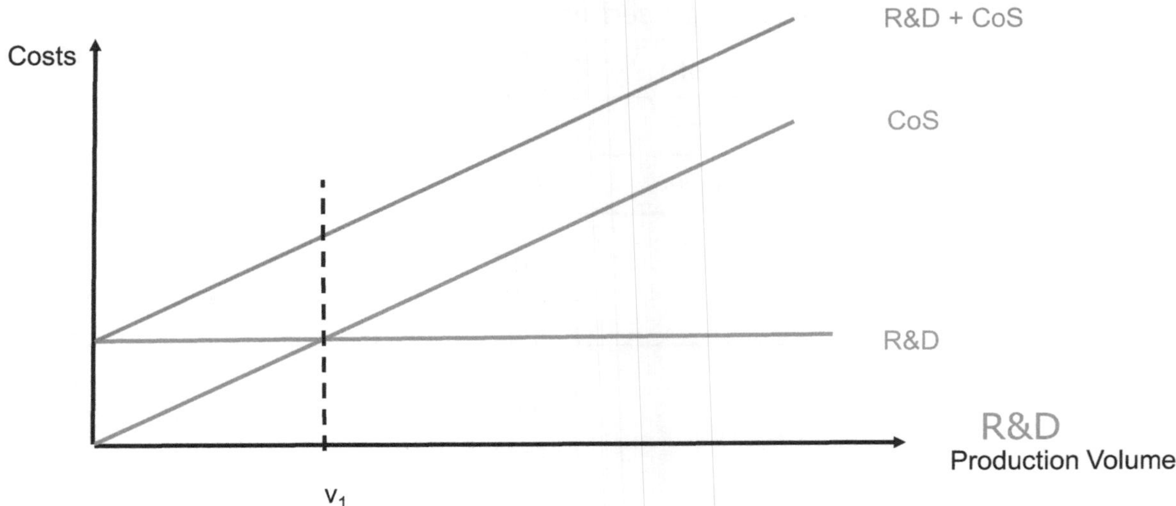

For any product development projected (and actual) lifetime volume is important input for overall cost optimization. For a low volume product v<v1 in graphic above lifetime cost for R&D are higher than for CoS. Looking at costs only R&D portion is right area for optimization. For semiconductor development salaries for engineering, mask set for production and cost for software and IP licenses are typical contributors to these one-time costs.

For product volumes with v>v1 CoS portions dominate. Invest into additional R&D to optimize these costs pays-off if you get more good dies from each wafer. This can be realized by less area for a single die and/or better yield for single wafers. Optimizing other manufacturing costs like test or package costs may also rectify additional R&D investments.

Please note, there might be other reasons for investment into R&D. Time-to-market or other strategic goals can move cost considerations to lower priorities.

IJTAG data formats ICL and PDL

The following summary on data formats ICL and PDL as defined by IEEE 1687 is provided with [43]. Please note, the term 'scan path' in the context of IEEE 1687 is used to describe a path from a test access port to test data registers of specific instruments.

Instrument Connectivity Language (ICL)

The description language ICL introduced with iJTAG is used to describe the architecture of the Instrument Access Network. Its structure is quite similar to that of a hierarchical netlist. However, ICL does not describe the behavior of the contained elements but only their interfaces and the connections to each other. An ICL description is composed of at least one or multiple individual ICL modules. Individual ICL modules can be instantiated within each other to create a hierarchical structure. The ICL module at the top of the hierarchy represents the top-level module. An ICL network is an ICL description that may contain one or more controllers (e.g., TAP controller) on one side and at least one instrumentation on the other side. Each ICL module has an own description specifying its interface and building blocks. A building block of an ICL module may be components like scan/data registers, ports with specific functions, scan/data multiplexers or ICL instances of other modules. The iJTAG standard defines several ICL port functions. These port identifiers are used to identify the specific functions of the related ports. For example, the port functions DataInPort or DataOutPort define ports that are intended for reading and writing test data. The CaptureEnPort, ShiftEnPort and UpdateEnPort functions are used to define ports that receive the respective control signals from the TAP controller. Furthermore, the port identifiers ClockPort and ResetPort define the ports which are explicitly intended for clock and reset signals. In addition to other port functions, an ICL module can also contain scan or data multiplexers. Scan multiplexers are used to make scan paths reconfigurable and form the basis of a SIB. Data multiplexers are used, for example, to select whether functional data or test data is present at the module's ports. With data multiplexers it is thus possible to switch between the normal operation or the test operation of the module. iJTAG TDR of desired length are built based on the ICL building block ScanRegister. Each element in ICL is identifiable in the network by its instance path. This instance path indicates the localization of the instance in the network according to the hierarchy starting from the top-level instance.

The description of the iJTAG network in ICL is necessary to enable so-called retargeting. This means that the active scan path is adapted to a test procedure defined in PDL in such a way that the instrumentations intended for the test procedures are identified and located on it. To realize test procedures in iJTAG, another description language is required in addition to ICL, namely PDL.

Procedural Description Language (PDL)

PDL is used to describe the operations to be performed on the instruments. In conjunction with the ICL description of an instrument, it is possible, for example, to define read and write operations in terms of stimuli and expected responses to the ICL registers using PDL. To access the ICL registers it is important to know the ICL network description. Only then PDL procedures can be mapped to the corresponding TDR of the target module through the hierarchy of the ICL network. PDL is defined by a set of commands that can be divided into two levels. PDL level-0 includes the basic commands to describe test procedures. PDL level-1 includes all commands from level-0 and extends them with a few additional commands. The level-1 PDL commands are only relevant in interactive silicon debug sessions. They are not relevant for the generation of traditional manufacturing patterns. However, PDL level-1 allows the integration of the scripting language Tool Command Language (TCL). This integration of TCL makes it possible to extend the functionality of PDL to that of a programming language. For example, loops, conditions, or variables can be used to make the description of test procedures much more flexible and complex. All PDL commands of level-0 and level-1 are fully TCL compatible and integrable. The PDL level-0 commands can be divided into two classes of commands: setup commands and action commands. Setup commands initially have no direct effect on the instrument, they are queued until executed by an action command. An action command has direct influence on the instrument by executing the queued setup commands. Table below gives an overview of all PDL level-0 commands. The most important setup commands are iWrite, iRead and iScan. Listing below shows an example of a PDL script. iWrite is used to write a specific value to an ICL register or port. Therefore, a valid ICL port or register and the value to be written must be given as arguments to an iWrite command. For example, line 9 of Listing below defines a write operation of the value '0b1' to the ICL register BIST_start. An iWrite command thus includes shifting test data to the TDR bits associated with the ICL port or register and then passing the test data by an update operation. With iRead the current value of an ICL port or register is read and compared with an expected value (see Listing below line 14). This means that following an iRead command, the data is first captured into the TDR bits of the ICL port or register by a capture operation and then the data is shifted and read out via TDO. An iScan command combines an iWrite and iRead command and thus performs capture, shift as well as update operations. Other important setup commands are iCall and iMerge. iCall can be used to call and execute already defined iProc procedures. In the arguments of the iCall command, the required arguments of the called iProc procedure are passed. The use of iMerge allows parallel execution of iCalls. Between the commands iMerge -begin and iMerge -end any number of procedures can be called using the iCall Command. The retargeting tool then tries to generate test patterns that allow the best possible parallel execution of these procedures. However, setup commands are first put into a queue and are only executed when they are triggered by an action command. The most important action command is iApply. This command is explicitly used to execute the previously defined setup commands. The set of previously defined and queued setup commands followed by an iApply command is called an iApply group. The queued setup commands of an iApply group are executed in a non-deterministic order. However, if the same port or register is written or read several times within an iApply group, the last defined value always dominates. An iApply group is then converted to test patterns as well as the required capture, shift and update operations. Another important action command is iRunLoop. It allows to define wait sequences. Either a time or a number of cycles of a given clock can be specified as an argument to this command (see Listing below line 12).

```
1 iProcsforModul e Instrument_A
2 iProc start_bist {bist_mode SINGLE} { test_value 0b010 } {
3
4        if {${bist_mode} == MULTIPLE} {
5                iWrite BIST_mode [1:0] 0b11
6        }
7
8        iWrite BIST_init [ 2 : 0 ] $test_value
9        iWrite BIST_start 0b1
10       iApply
11
12       iRunLoop 1000 -tck
13
14       iRead BIST_result [2:0] 0xe
15       iApply
16 }
```
Listing: Sample PDL Script

A PDL script starts with referencing the commands and procedures defined in the script to an ICL instrumentation using the command iProcsForModule. The PDL script in Listing above is defined for Instrument_A. After referencing, the procedure start_bist is defined for the instrument_A using the command iProc. Two arguments are passed to the start_bist procedure, which can be used as the bist_mode and test_value variables in the procedure. The values in the braces reflect the default values for the arguments that are to be used if the arguments are not initiated. When the argument MULTIPLE is passed to the procedure, the port BIST_mode[1:0] is also written in addition to the ports BIST_init[2:0] and BIST_start. After that, a wait sequence of 1000 test clock cycles is executed. Finally, the test results are read out at the port BIST_result[2:0] and compared with the expected result '0xe'.

Command	Type	Purpose
iPDLLevel	Setup	Identify PDL flavor
iPrefix	Setup	Specify Hierarchical prefix
iReset	Action	Reset The network

iWrite	Setup	Queue data to be written
iRead	Setup	Queue data to be read
iScan	Setup	Queue data to be scanned
iOverrideScanInterface	Setup	Indicate the capture, update and broadcast behavior to be imposed on a list of scan interfaces
iApply	Action	Execute queued operations
iClock	Setup	Specify a system clock, which is require to be running
iClockOverride	Setup	Override definition of system clock when it is generated on-chip
iRunLoop	Action	Issue a number of clock cycles
iProc	Setup	Wrapper for a PDL procedure
iProcsForModule	Setup	Identify the module in the ICL with which subsequent iProcs are associated
iUseProcNameSpace	Setup	Use namespace for subsequent iCalls
iCall	Setup	Invokea PDL procedure
iNote	Action	Send text to runtime
iMerge	Setup	Allow merging (concurrent evaluation) of iCalls
iTake	Setup	Disallow other merge threads from modifying a model resource
iRelease	Setup	Re-allow other merge threads to modify a model resource
iState	Action	Document the current state of the network

To execute the setup commands of an iApply group, it is necessary that the IJTAG Instrument Access Network is configured in such a way that the respective target instrumentation is located on the active scan path. The transformation of PDL operations into corresponding test patterns for configuring the network is covered in more detail in the next section.

PDL retargeting

To ensure the reconfigurability of the iJTAG Instrument Access Network, it is necessary that the active scan path is adjusted according to the intended test operations. Thereby it is important that not only the intended instruments and scan segments are on the active scan path, but also that the network is configured in such a way that the access time to the instruments is minimized. The process of translating the commands of an iApply group into corresponding scan vectors to configure the Instrument Access Network is called retargeting. A retargeting step must perform two tasks. It must transform PDL sequences for an instrument from the IP level to a higher level (top level) of the design to generate the necessary capture, shift and update operations to make the instrument part of the active scan path. And it must translate the read and/or write operations on the instrument into corresponding scan vectors that perform the read and/or write operations.

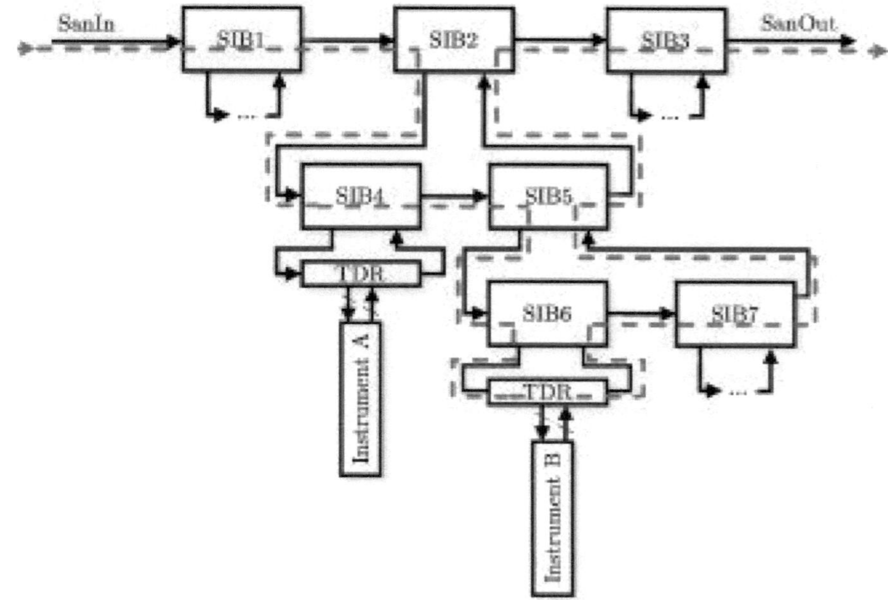

Figure 99: Sample retargeting process

For example, to perform an iWrite or iRead command to write to or read from a register of Instrument B in Figure 99, the SIBs must be configured such that the TDR of Instrument B is part of the active scan path. A retargeting tool must therefore first configure the scan vector to close SIBs 1, 3, 4 and 7 (if only the TDR bits of Instrument B should be on the active scan path) and open SIBs 2, 5 and 6. The resulting active scan path is highlighted by dashed lines in Figure 99. Then the iWrite and/or iRead commands are translated and added to the scan vector. The final scan vector can subsequently be inserted via ScanIn. A retargeting process is performed by an EDA software tool using ICL and PDL descriptions as input.

Discussing about reliability touches also similar topics like availability and safety. Here the terms are differentiated as follows:

- Reliability is the continuity of a correct service
- Availability is the readiness for a correct service
- Safety is the absence of catastrophic consequences on the user and the environment (see Chapter Functional safety and DfT for details)

A prominent example for availability is the up time of a telecom network or of a radio station. These are typically in the range of years. Such a time interval also includes maintenance activities and the exchange of failed components which are not impacting the overall service provision.

A clear separation of the terms fault, error and failure is necessary:

- Fault is an abnormal condition that *can* cause an element to fail.
- Error is a discrepancy of results which *are* computed, observed, or measured.
- Failure is the termination of the ability to *perform* the required function.

A failure within a component could be a fault at the next level, e.g. on system. To avoid that errors result in failures redundancy measures can be in place. A prominent example are the use of error-correcting codes (ECC).

The following metrics and terminology are taken from [44].

Reliability R(t) is the probability as a function of time that the system has been up continuously in the time interval [0, t]. Availability A(t) is the average fraction of time over the time interval [0, t] that the system is up.

For reliability three parameters are defined:

- Mean time to failure (MTTF)
- Mean time between failures (MTBF)
- Mean time to repair (MTTR)

Using these parameters long-term availability A can be defined as:

$$A = \text{MTTF/MTBF} = \text{MTTF/(MTTF+MTTR)} \text{ with} \lim_{t \to \infty} A(t) \quad .$$

Functional safety and DfT

Functional safety requires that the malfunctions of electrical and electronic systems should not cause harm to human beings. The basics of functional safety (FuSa) are presented in a tutorial paper [45]. This chapter summarizes the main concepts, terminology and metrics based on the mentioned tutorial.

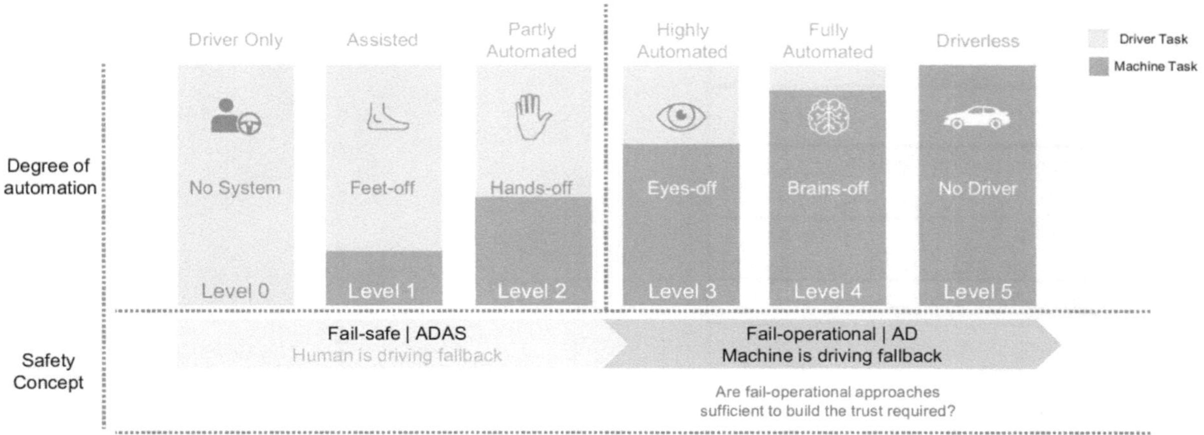

Figure 100: ADAS Levels (Source: Barclays Research & Infineon)

As shown in Figure 100 there are six levels for Advanced Driver Assistance Systems (ADAS) defined. With increasing levels, the role of the driver is vanishing, and the machine is taking over more control of the car. From a safety perspective a fail-safe behavior is required for level 0 to level 2 (or 2.5). The system transits to a known safe state so that a failure could not cause any harm. For the higher ADAS levels a fail-operational state is required. Here the system can continue operation in presence of component failures.

A safe vehicle comprises four elements:

1. Road safety: reducing accidents by human errors
2. Device reliability: zero failures of the device. The goal is to improve the manufacturing quality and the design robustness to reduce the failure rates of the components.
3. Functional Safety: zero accidents by failures of systems in automobiles
4. Security (or Cyber Security): resilience against system hacks

ISO standard 26262 defines Functional Safety (FuSa) for E/E systems[9] on road vehicles which is the focus of this chapter. FuSa according to ISO 26262 is addressing faults or failures. Another concept is SOTIF (Safety of the intended function). Here safety is impacted by a used functionality which is of limited performance for the intended safety function. The presence of faults is not considered for SOTIF.

---

[9] Electric/electronic systems

Reliability influences FuSa since failure rates could influence safety measures. Availability is not always necessary for maintaining safety of the vehicle. E.g. key-on tests detect a fault and the system will not start the engine. This reduces availability but guarantees FuSa.

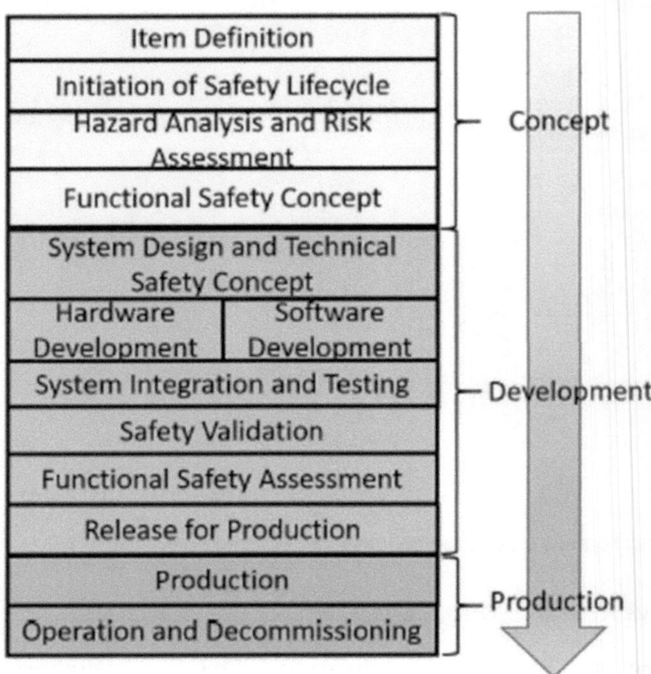

Figure 101: Lifecycle of safety-related products

The lifecycle of safety-related products according to ISO 26262 is shown in Figure 101 (taken from [45]). The item for safety analysis is typically defined by the car manufacturer[10]. An item is a function or part of a function on vehicle level. The hazard analysis and risk analysis for the item has the goal to quantify the risk of hazards caused by failures of the functions. Malfunction has two aspects. These are not correctly executed functions when required or execution when not required. An example for the second aspect is the ignition of an airbag during normal drive. Operating conditions and driving scenarios where malfunctions can occur are considered for hazardous events. The risk is assessed based on three factors:

1. Severity (how much harm?)
2. Exposure (how often it's likely to happen?)
3. Controllability (what is the likelihood that the hazard can be controlled?)

---

[10] OEM (Original Equipment Manufacturer), used in automotive supply chain as synonym for car manufacturer. Suppliers are called Tier-1 and Tier-2. Tier-2 includes semiconductors.

Based on the risk level four Automotive Safety Integrity Levels (ASIL) can be defined. These are named ASIL A, B, C and D. The classification is executed for a certain function (e.g., Airbag) and results in the safety goal for that function (e.g., ASIL-D). Additionally, there is a class QM without any safety requirements to comply with.

FuSa for automotive SoC does not affect all parts of the lifecycle from Figure 101. The concept of faults is important to quantify the risks. The categorization scheme is shown in Figure 102 which is taken from [45].

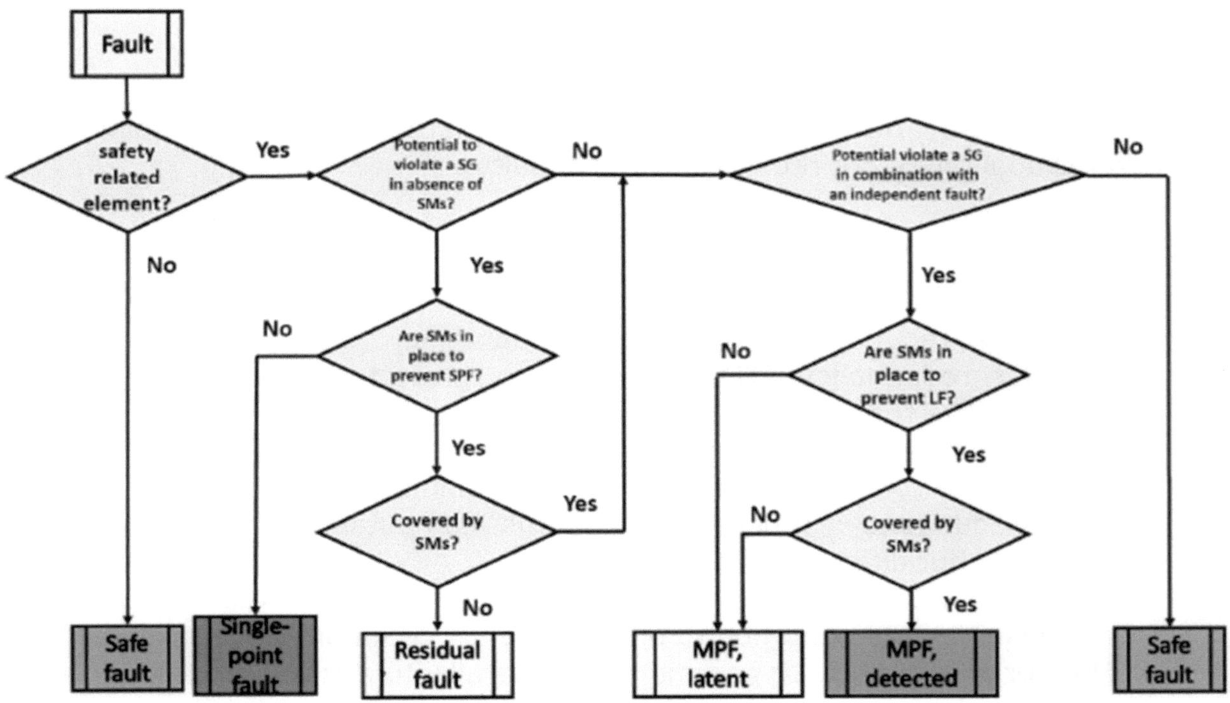

Figure 102: Fault categorization according ISO 26262

The fault categorization starts with the assumption of a 'fault' and considers the safety goal (SG) and available safety mechanisms (SM). Safe faults (left in Figure 102) can be excluded from the safety analysis. Single-point faults (SPF) violate a safety goal and no safety mechanism is in place. If a safety mechanism is in place but the fault is not covered by it, this is called a residual fault. For latent multiple-point faults (latent MPF or latent) there is no safety mechanism in place, or the safety mechanism is unable to cover it. If a safety mechanism is available and able to cover the MPF, this is classified as MPF detected. MPF with order greater than 2 are classified as safe faults since more than two occurrences are regarded as unlikely.

Based on the fault categorization above diagnostic coverage can be determined by FMEDA (failure modes, effects, and diagnostic analysis). Single-point faults and latent faults are the two metrics of diagnostic coverage. The raw failure of a safety-related hardware element is defined as λ.

$$\lambda = \lambda_{SPF} + \lambda_{RF} + \lambda_{MPF,D} + \lambda_{MPF,L} + \lambda_{S}$$

The raw failure λ is the sum of all faults categorized according Figure 102.

The **single-point metric** is defined using the sum of single-point faults and residual faults for all safety-related hardware elements as follows:

$$1 - \frac{\sum\limits_{HW,SR} \lambda_{MPF,L}}{\sum\limits_{HW,SR} \lambda_{HW,SR} - \lambda_{SPF} - \lambda_{RF}}$$

The **latent fault metric** defines the portion of latent MPF if single-point faults and residual faults are removed from all faults:

Safety mechanisms are categorized by error detection methods:

- Hardware redundancy: dual-mode redundancy like lockstep or triple-modular redundancy (TMR) is used to protect SoC-level e.g. critical registers like trimming values (implemented as Triple Voting flip-flops)
- Information redundancy: error-correction codes and parity check, typically protect data communication channels and memories
- Time redundancy: repeating the same computation, could use same hardware but different algorithms
- Redundancy used in combination with diversity, can be in time, algorithms or physical implementation to protect against common-mode failures
- Monitor critical parts or parameters (in parallel to mission mode)
- Test (during in-active phases of mission mode): LBIST, MBIST or SW-based BIST routines, also: loopback for radar interface (MMIC)

Test-based mechanisms enabled by DfT (LBIST, MBIST) provide protection against latent faults. To cover single-point faults by BIST methods the Fault Tolerance Time Interval (FTTI) has to be shorter than the BIST execution including initialization and checking results. For applications with cyclic offline intervals this is an option, e.g., for radar functions. Typically, to cover single-point faults errors need to be detected by redundancy or continuous monitoring.

Based on the risk classification according to severity, exposure, and controllability a function is assigned an ASIL class. Some examples for a car are shown in Figure 103, taken from [46].

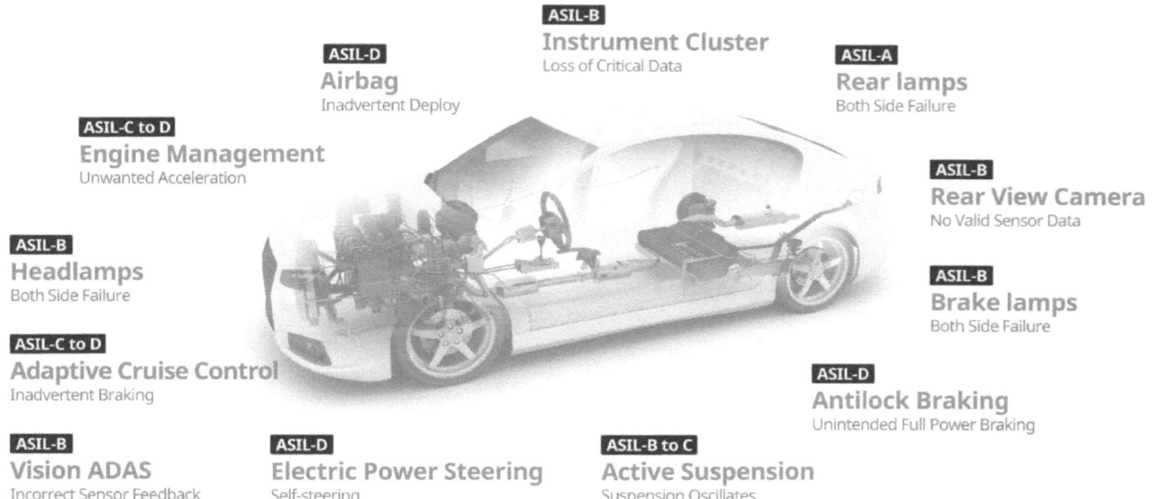

Figure 103: Examples for ASIL classifications

Some requirements for the ASIL classes are given with the following table:

	ASIL B	ASIL C	ASIL D
Single-point fault metric	90%	97%	99%
Latent fault metric	60%	80%	90%
Failure rate / h	$< 10^{-7}$	$< 10^{-7}$	$< 10^{-8}$
FIT [10^{-9} h]	< 100	< 100	< 10

The last two lines in the table above give Probabilistic Metric of Random Hardware Faults (PMHF). I.e., the safety mechanisms are already applied resulting in the listed failure rates or FIT numbers.